Getting Started with 3D Printing

A Hands-on Guide to the Hardware, Software, and Services Behind the New Manufacturing Revolution

Liza Wallach Kloski and Nick Kloski, Cofounders of HoneyPoint3D™

Getting Started with 3D Printing

by Liza Wallach Kloski and Nick Kloski

Printed in Canada.

Published by Maker Media, Inc., 1160 Battery Street East, Suite 125, San Francisco, CA 94111.

Maker Media books may be purchased for educational, business, or sales promotional use. Online editions are also available for most titles (*http://safaribooksonline.com*). For more information, contact O'Reilly Media's institutional sales department: 800-998-9938 or *corporate@oreilly.com*.

Editor: Roger Stewart
Technical Reviewer: Matt Stultz
Production Editor: Colleen Cole
Copyeditor: Kim Cofer
Proofreader: Charles Roumeliotis
Indexer: WordCo Indexing Services
Interior Designer: David Futato
Cover Designer: Karen Montgomery
Cover Illustration: HoneyPoint3D™
Illustrator: Rebecca Demarest

May 2016: First Edition

Revision History for the First Edition

2016-04-27: First Release

See *http://oreilly.com/catalog/errata.csp?isbn=9781680450200* for release details.

978-1-680-45020-0

[TI]

Contents

Part I Applications of 3D Printing

Part II Hardware and Printing Choices

Part III CAD Tutorials

Part IV The Future

Foreword

If you're reading this book, that means that you're about to embark on a potentially life-changing journey.

There's a good chance that a 3D printer will wind up in your home, if it hasn't already, and you're going to need to know how to use it.

Getting Started with 3D Printing can be your guide through the trials and tribulations of the 3D printing learning curve.

When you begin, you may not know a thing about this game-changing technology, but as you follow along with this book, you will be initiated into an elite society of alchemists who are capable of transforming the very matter of the physical world around you. Like me, you will become a maker.

Makers date back tens of thousands of years. While Neanderthals were roaming around the plains of the Serengeti, dragging their knuckles through the dirt, we makers were learning to transform sticks, stones, and animal bones into tools with which to hunt and survive. By the time some made rocks with handles, we makers had already moved on to bronze, smelting spearheads, and proto-writing our names into mountaintops.

The times have changed, but our powers have not.

The newest generation of makers wield the tools of digital fabrication to transmute thought into reality. The CNC machine carves metal, the laser cutter etches wood, and the PCB writer harnesses the power of electricity to drive even the most complex of robotics. But there may be no tool of alchemy more powerful than the 3D printer.

I can tell you from personal experience that learning how to 3D print may not always be fun, but it will always be rewarding. I remember watching my first test cube begin printing, and it was nothing short of miraculous. I had found the digital model on

Thingiverse and, all of a sudden, it was materializing before my eyes.

Life, when you love 3D printing, can be a tumultuous one of clogged print heads, broken belt drives, and unruly 3D CAD models. There is hope, but you have to start somewhere. And there's no better place to start than with this book, written by two educational alchemists—Liza Wallach Kloski and Nick Kloski of HoneyPoint3D™.

They have already empowered more than 6,000 apprentices, having successfully launched their own wildly popular educational program devoted to teaching the ins and outs of 3D printing. Liza and Nick will be able to accompany you on your journey as two helpful guides who will be there for you when your wires are crossed, your extruders are clogged, and your bed isn't level.

Soon, you will understand how to turn the mental into the physical, the digital into the material. You'll be able to clone everyday items, teleport them across thousands of miles, or reconfigure them to suit your own needs. You'll befriend other mystics, like yourself, who can help you to hone your powers. You might even join a local chapter of the elders, a makerspace, where secret rites and rituals will see you unite your powers with the collective to unleash even mightier projects on the world at large. And as the technology grows more powerful as a result, you'll be able to fabricate almost anything, practically out of thin air, moving from full color sculptures to complete electronic devices!

That is what you can expect when you end your journey. Once you complete this guide, you, my friend, will become one of the initiated. You will be a maker.

—Michael Molitch-Hou
Editor-in-Chief
3D Printing Industry (3DPI)
The leading news site dedicated to 3D printing

An Introduction to 3D Printing

Feel fortunate! You are living at a time when technology is increasingly helping people become masters of their environments. A 3D printer puts the power of a manufacturing plant on your desk and opens worlds of opportunities that you (and the rest of humanity) have never experienced before.

Industrial 3D printing has been around since the 1980s. The technology became available to hobbyists and consumers in 2009 when the RepRap project brought together thinkers and coders from around the world to create a freely open codebase. This gave anyone the ability to build a personal 3D printer, such as the very early example shown in Figure P-1.

Using an open source design (which grants a free license to a product's blueprints), dedicated hobbyists were able to affordably access what was otherwise a very expensive technology. This marked a cornerstone in the ability of ordinary customers to access desktop 3D printing.

Fast-forward to 2016. The 3D printing industry (industrial and consumer combined) is now estimated at around $7 billion and is expected to exceed $17.2 billion by 2020. No one has a crystal ball to know the future, but it seems certain that 3D printing will change the way we as a society design and manufacture physical items.

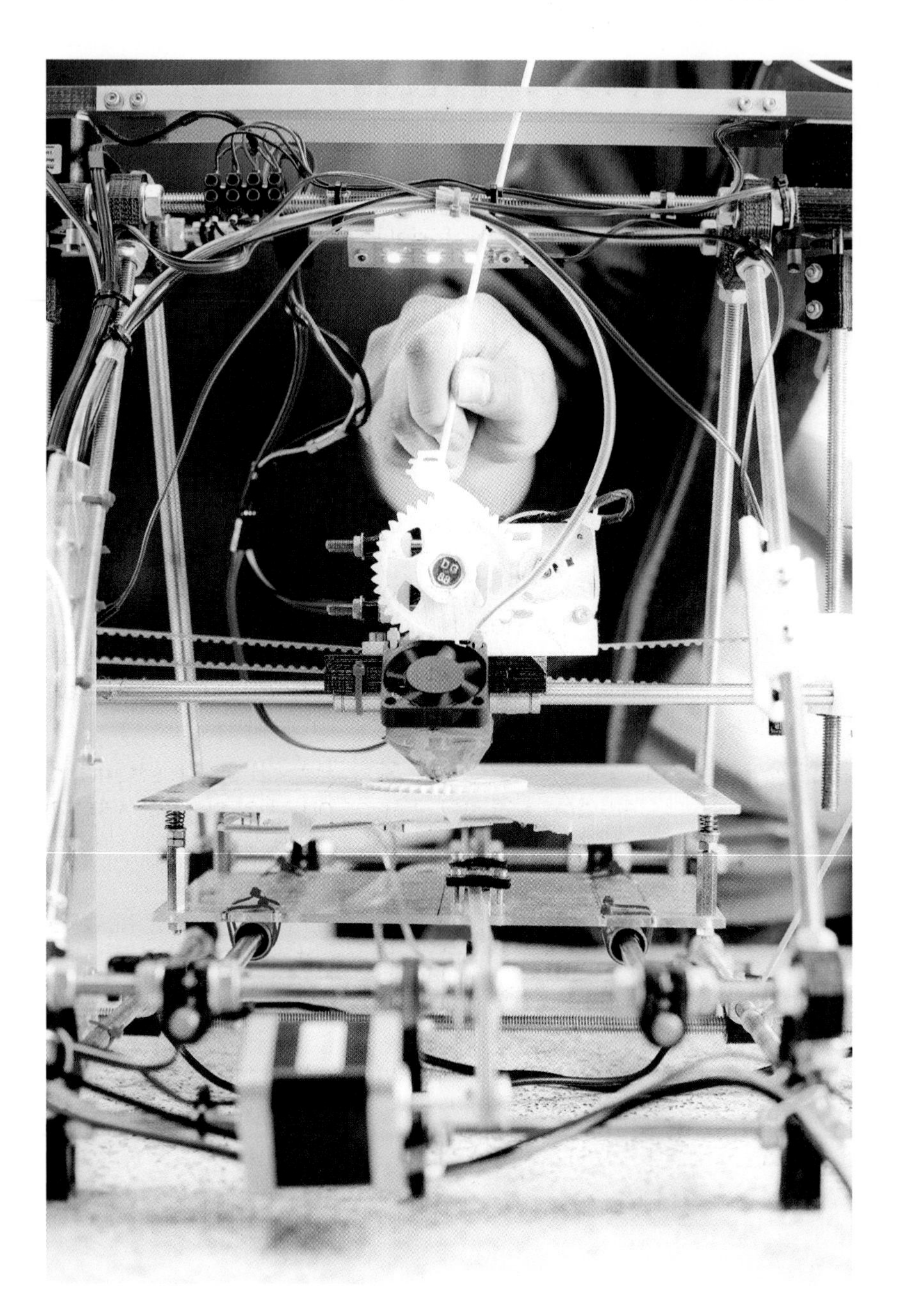

Figure P-1. *An early example of an assembled open source 3D printer kit*

How 3D Printing Works

The most common type of consumer 3D printing, also known as "additive manufacturing," is called FDM (fused deposition modeling) or sometimes FFF (fused filament fabrication). FDM printers create objects by adding material in successive layers, building up an object layer by layer over time. A thin strand of filament feeds into a part of the machine called an extruder, which melts the plastic at a high temperature—typically around 200°C. There are many other methods of 3D printing that this book will cover in later chapters, and they all work in the same "additive" way of creating an object by adding material layer by layer.

The most recognizable tool you can compare this process to is a hot glue gun. You probably have one at home and have used it for craft or school projects. When you squeeze the handle, the glue is pressed against a heating element and soft, spaghetti-like strands extrude from the nozzle. Imagine swirling those strands around and around on top of each other, forming circles that build up into a tube. As the glue cools, it hardens.

Following the glue gun analogy, in a 3D printer the thin strand of melted plastic is laid down, layer by layer, on a flat surface where it cools and hardens into an object. The image in Figure P-2 provides a closer look at how this technology works.

Most consumer 3D printers use FDM technology and have a large spool of coiled plastic called filament attached to them. (We will talk about other types of 3D printer technologies in later chapters.)

The 3D printer knows precisely where to trace each layer through instructions in a digital file sent from a computer.

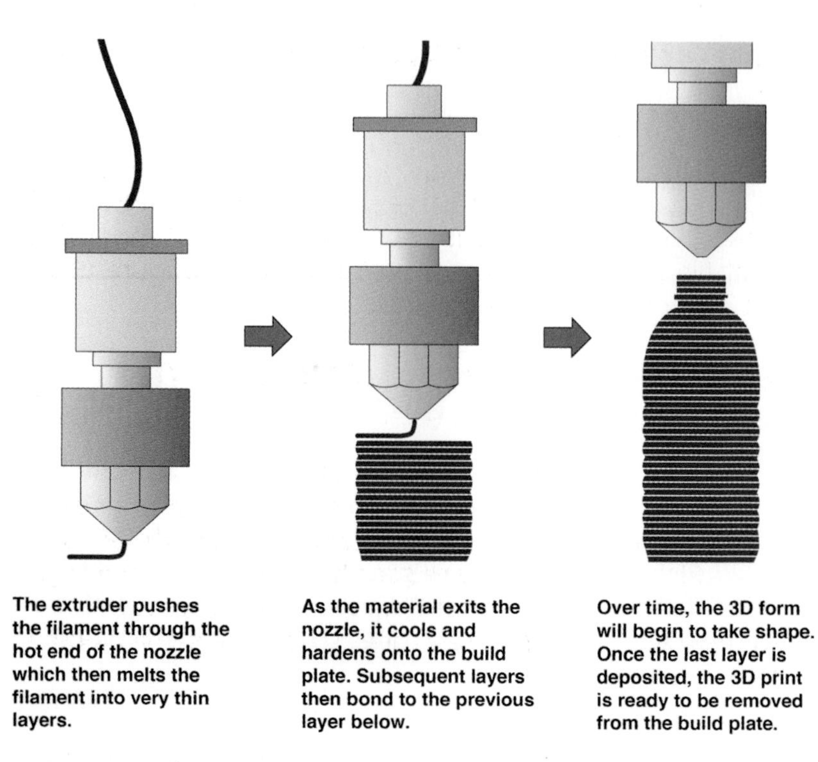

The extruder pushes the filament through the hot end of the nozzle which then melts the filament into very thin layers.
As the material exits the nozzle, it cools and hardens onto the build plate. Subsequent layers then bond to the previous layer below.
Over time, the 3D form will begin to take shape. Once the last layer is deposited, the 3D print is ready to be removed from the build plate.

Figure P-2. *An extruder is depositing melted plastic one layer at a time to build a bottle (infographic by Jeff Hansen, Honey-Point3D™)*

3D Printing Is Not Like 2D Printing

In 2013, at our retail 3D printing store in Oakland, California, we used to get a lot of questions from casual visitors along the lines of, "Where do you put the ink?" and "How much paper does it need?" The majority of the public hadn't made the connection that 3D printers (which print three-dimensional objects) are *very* different from 2D printers (which print flat images on paper).

Today, ninety percent of our store visitors—from all age groups, backgrounds, and interests—know the difference between 2D printing and 3D printing. 3D printing has received a tremendous amount of media attention these past few years, and the general public has made a huge leap in awareness of the technology.

3D printing is, therefore, a paradigm shift from our traditional understanding of printing based on paper-and-ink printers. You can't print a "flat" graphical image such as a PDF or JPEG in 3D. A special kind of digital file is needed to make a 3D print. That type of file is called a *3D model*. We will go into more details about 3D models later in this book.

Managing Expectations as You Start Your Journey

3D printing is more difficult than it should be. The technology is still relatively young, and the machines are not as plug-and-play as most home appliances, like a microwave oven. Expect to go through a learning curve where good old trial-and-error will give you lots of valuable experience and, eventually, success! You will learn the tricks of the trade as you go. Here are some common experiences you should expect:

Prints fail
: A lot. Especially when you are first learning. Did you set the wrong temperature? Did the 3D file have errors in it? Was the print bed not level? You'll be asking yourself these questions and more as you experiment.

Prints take a long time
: Want to print a phone case? No problem! But the print will likely take 3+ hours.

Printers need ongoing maintenance
: Motor belts pop off. The hot end gets jammed. Stuff happens and, in most cases, it will be *you* fixing it.

Sometimes prints need pre/postprocessing
: Some 3D model files you find on the Internet can't be printed as-is and will need to be fixed. Printed 3D objects will need work to get the smooth surface you may desire. (You will find that sandpaper does wonders with those rough edges on a newly hatched 3D print.)

Don't Worry! This Is Why We Wrote the Book

3D printing isn't just a technology, it's an ecosystem of software and hardware. Implementing 3D printing takes a three-pronged approach: education, CAD file generation, and the physical aspect of 3D printing. You'll get a basic grasp of this system by reading this book. The rewards of learning how to 3D print are well worth the learning curve, and you'll be glad you made the effort. Reading this book will save you time, money, headaches, and heartaches and will lessen the number of "valuable lessons" you have to go through on your own. We made plenty of mistakes so *you* won't have to!

What's in This Book?

Getting Started with 3D Printing offers you a clear roadmap of best practices to help you successfully bring 3D printing into your home, classroom, or workplace. We hope you will find it a fun, practical guide that will help you navigate your way from initial curiosity to active, hands-on 3D printing. The book is intended for those who have no prior experience in 3D printing. Even if you do have some experience with 3D printing, however, you may find that the book will help you see a bigger picture and greater possibilities.

This book will inform you about:

- How 3D printers work
- What to look for when buying a 3D printer and supplies
- Setting up and maintaining your own 3D printer
- Outsourcing 3D modeling and printing services
- Creating and fixing your own 3D models
- Designing a personal 3D printing makerspace
- The future of 3D printing

Fasten your seatbelt and prepare yourself for this fast-approaching "manufacturing revolution."

It may just change your life!

Safari® Books Online

Safari Books Online is an on-demand digital library that delivers expert content in both book and video format from the world's leading authors in technology and business.

Technology professionals, software developers, web designers, and business and creative professionals use Safari Books Online as their primary resource for research, problem solving, learning, and certification training.

Safari Books Online offers a range of plans and pricing for enterprise, government, education, and individuals.

Members have access to thousands of books, training videos, and prepublication manuscripts in one fully searchable database from publishers like Maker Media, O'Reilly Media, Prentice Hall Professional, Addison-Wesley Professional, Microsoft Press, Sams, Que, Peachpit Press, Focal Press, Cisco Press, John Wiley & Sons, Syngress, Morgan Kaufmann, IBM Redbooks, Packt, Adobe Press, FT Press, Apress, Manning, New Riders, McGraw-Hill, Jones & Bartlett, Course Technology, and hundreds more. For more information about Safari Books Online, please visit us online.

How to Contact Us

Please address comments and questions concerning this book to the publisher:

Maker Media, Inc.
1160 Battery Street East, Suite 125
San Francisco, California 94111
800-998-9938 (in the United States or Canada)
http://makermedia.com/contact-us

Make: unites, inspires, informs, and entertains a growing community of resourceful people who undertake amazing projects in their backyards, basements, and garages. Make: celebrates your right to tweak, hack, and bend any technology to your will. The Make: audience continues to be a growing culture and community that believes in bettering ourselves, our environment, our educational system—our entire world. This is much more than an audience, it's a worldwide movement that Make: is leading—we call it the Maker Movement.

For more information about Make:, visit us online:

Make: magazine: *http://makezine.com/magazine*
Maker Faire: *http://makerfaire.com*
Makezine.com: *http://makezine.com*
Maker Shed: *http://makershed.com*

We have a web page for this book, where we list errata, examples, and any additional information. You can access this page at: *http://bit.ly/getting-started-with-3d-printing*

To comment or ask technical questions about this book, send email to *bookquestions@oreilly.com*.

Acknowledgments

It has been an honor to work with Maker Media and to write about a subject we believe will change the world. We would like to thank the following people and companies that supported our process and research, and gave us endless amounts of encouragement in writing this book:

Roger Stewart
Brian Jepson
Neil Edde
Alicia Moszee
Jeff Hansen
Susan Wallach
Deborah Wallach
Ari Wallach
Autodesk
Kudo3D
Printrbot
Type A Machines
Solid Professor
Breathe-3DP

PART I

Applications of 3D Printing

1/You Say You Want a Revolution?

There is no doubt that the rise of 3D printing is laying the foundation for a significant change in our society, one that will have important echoes in the coming years and decades. This chapter gives you a short historical perspective on the manufacturing roots of 3D printing and how it currently affects our lives.

The rise of 3D printing is being heralded by some as the beginning of a "third industrial revolution," but we feel it is better called the "personal manufacturing revolution"—a term attributed to Avi Reichental, former CEO & President of 3D Systems.

The first and second industrial revolutions developed new manufacturing processes that included going from hand production methods to machines, new chemical manufacturing processes, increased use of steam power, the creation of machine tools, and mass production (Figure 1-1). The second industrial revolution, which introduced assembly lines, continues to affect how we manufacture goods today.

These two revolutions marked major turning points in history; almost every aspect of daily life was influenced in some way. Average incomes and population numbers began to exhibit unprecedented growth, and the standard of living rose for most people.

3D printing, like its fabrication predecessors, has created new manufacturing processes, and it promises to change not just manufacturing but our way of life. The key difference is that *this* new manufacturing technology is very personal. Essentially, it is a factory on your desk, where you can make almost anything you wish. You can manifest your ideas in ways that were once only available through industrial prototyping. You don't need permission from a board of directors or even orders from customers to produce new products. You just need your imagination.

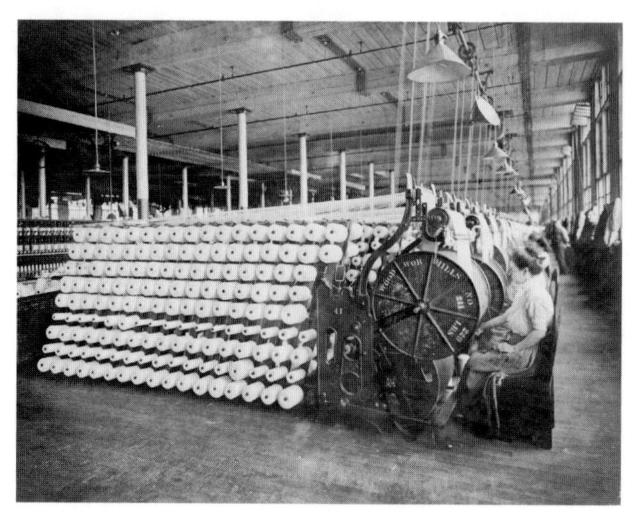

Figure 1-1. *Woman working in yarn mill, showing mass production circa 1900*

This is why we and many others believe "personal manufacturing revolution" is a more fitting term for the changes 3D printing will bring. It shifts the focus to the individual, showcasing self-expression and self-sufficiency.

A Manufacturing Full Circle

3D printing won't completely replace traditional manufacturing processes. Rather, it will *augment* current means of mass production. In addition, the centers of manufacturing have come full circle with 3D printing. You can now manufacture either in rural or urban areas, production can move back from the factory to the small shop, and you can economically produce one piece or many. In this way, 3D printing has combined the best aspects of pre- and post-revolution manufacturing, as shown in Figure 1-2.

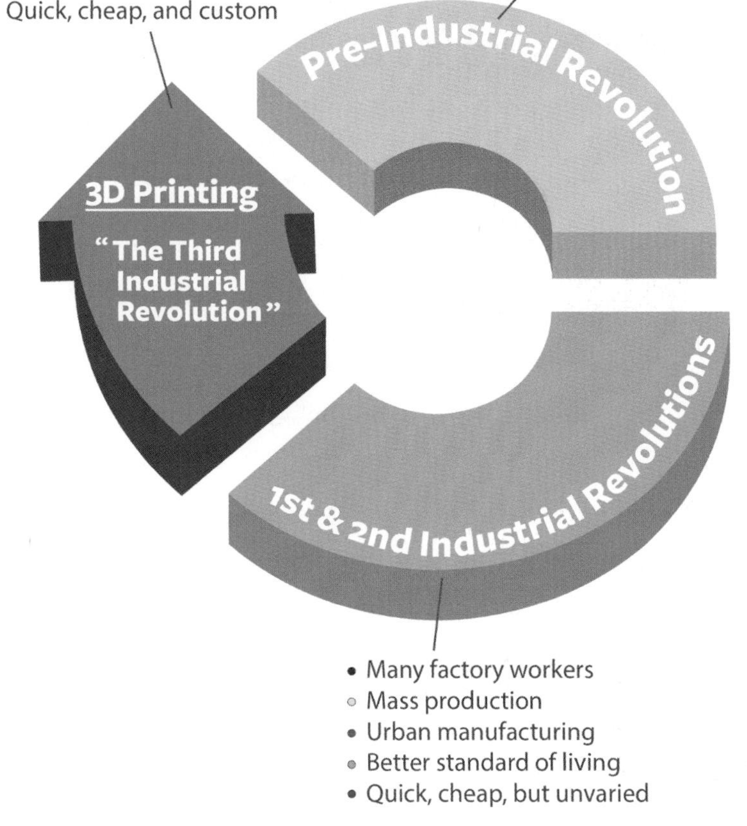
• Single worker production
• Low production volume
• Rural manufacturing
• Low standard of living
• Slow, expensive, but custom
• Scalable workforce
• Production of one, a few, OR many
• Not location dependent
• Even higher standard of living
• Quick, cheap, and custom
Pre-Industrial Revolution
3D Printing
"The Third Industrial Revolution"
1st & 2nd Industrial Revolutions
• Many factory workers
• Mass production
• Urban manufacturing
• Better standard of living
• Quick, cheap, but unvaried

Figure 1-2. *3D printing combines the best of all manufacturing periods and shows how manufacturing has come full circle (infographic by Jeff Hansen, HoneyPoint3D™)*

3D printing technology will inevitably change the world of manufacturing. Local design and production can be quick, efficient, and environmentally conscious. It can also produce highly customized products. Because each 3D print is made individually, modifications can be added between 3D prints (something that mass manufacturing cannot do easily).

In the next chapter, we'll examine how 3D printing is helping drive the maker movement.

2/3D Printing and the Maker Movement

Many people find the act of making very satisfying. New, inexpensive, and more readily available technologies are allowing makers and DIYers to do things they couldn't do before. Inexpensive 3D printers are enabling ordinary people to make their own jewelry, toys, mechanical parts, and many other items.

Impact of the Maker Movement

The "maker movement," an extension of the DIY community, has had a tremendous influence in promoting consumer awareness of 3D printing. Hundreds of thousands of people have learned about 3D printers and saw them in operation for the first time at Maker Faires. Figure 2-1 shows attendees at a local Mini Maker Faire listening to our lecture on 3D printing.

No matter what experience level you have, you can be a maker, too. And, as you will discover in this chapter, a whole ecosystem of companies, organizations, and services are available to help you use 3D printing to foster that ability in you.

Who Is a Maker?

Millions of people each year attend Maker Faires. So, who are these makers?

A maker is anyone who puts things together creatively. Look around you, makers are everywhere. The do-it-yourselfer making a wood door for his house is a maker. The hobbyist assembling a flying drone from a kit is a maker. A programmer developing an Arduino-based electronic device is a maker. They are all making something new by using their hands and creativity, turning their ideas into physical forms.

Figure 2-1. *Nick Kloski making a presentation on 3D printing at the East Bay Mini Maker Faire, Oakland, California, 2015*

In fact, you probably already are a maker in your daily life! Maybe something broke in your kitchen and you hacked something together to make it work again...you're a maker! The children in schools making their 7th grade science project...they are makers too! You don't have to have a garage full of tools to be a maker, you just need to think and build creatively.

Figure 2-2 shows some examples of how consumers, turned makers, are using 3D printing today.

As you tinker and experiment with this technology, you will undoubtedly find ways to use 3D printing to enhance your life. In fact, a study from Michigan Technological University estimated that a family can save anywhere from $300 to almost $2,000 a year by 3D printing items such as combs, cookie cutters, door stops, tool parts, and more. It's easy to see how 3D printing has found a happy home in the maker movement.

LEARNING
3D printing visual aides to supplement learning. 3D printing-centric course curriculum. Creation of after-school fab-labs.
PARENTING
Toys for children. A way for parents to engage with children in a maker-spirited environment.
HOME COOKING
Custom cookie cutters, ice cube molds, and other household kitchen items. In the future, consumers will be able to 3D print food at home.
How everyday makers use 3D printing
HOME REPAIRS
Replacement parts for appliances and other objects around the house, for example outlet covers, pieces for washer / dryer, doorstops, wall hooks.
HOBBIES
3D printing their own custom pieces for drone kits and various remote controlled vehicles & gadgets.
MEMORY KEEPSAKES
3D scanning and printing miniature self, family portraits, wedding cake toppers.

Figure 2-2. *Examples of how consumers, turned makers, are using 3D printing today (infographic by Jeff Hansen, Honey-Point3D™)*

How the 3D Printing Ecosystem Helps You Be a Maker

An entire ecosystem of products and services is being developed around this technology. Whether you are a maker, investor, business startup, or just testing the waters, it's important to be aware of the larger landscape of companies and services that make up the current 3D printing ecosystem.

Consumer-level 3D printing has only been around for a relatively short time, but the realm of related goods and services is growing very quickly. Figures 2-3 through 2-6 provide a list of company names as well as functional descriptions of the types of services and products you will encounter. You don't have to master all aspects of 3D printing yourself; you'll see that you can download ready-made 3D models, hire 3D design and CAD services, outsource the 3D printing to a service bureau, or purchase 3D printed objects. How you want to approach it is up to you!

The list of companies and services mentioned here is not comprehensive, but rather an overview of what you can expect to encounter in the 3D printing world. For a more comprehensive list, visit the online 3D Printing Directory (*http://directory.3dprintingindustry.com*).

In the next chapter we'll examine some of the ways that 3D printing is being used today to provide greater choice in consumer items, better health care, and more. You'll see that 3D printing is already having an important impact on our society.

3D Printer Manufacturers

There are over 150 consumer 3D printers on the market costing anywhere from $200 to $5000. Available as kits, or as fully assembled units, each printer has different specifications that make them unique (e.g., maximum build volume, types of materials you can print, maximum temperature, etc.)

Printrbot, Type A Machines, Kudo3D, FSL3D, Airwolf3D, Formlabs, Ultimaker, Afinia, Lulzbot, Reprap, 3D Systems, Stratasys, ZMorph, Makerbot, XYZ Printing, Gigabot, SeeMeCNC

Material Suppliers

All 3D printers require a material to print with, and you have a variety of materials to choose from. The printer hardware will determine what materials can be printed. Here is a list of popular material suppliers.

Breathe3DP, Taulman3D, Maker Juice, MadeSolid, ColorFabb, Reprap Austria, Asiga, MatterHackers, Proto-pasta

3D Software

In the realm of 3D printing, software is vital. From slicers to 3D modeling tools, you will be interacting with a piece of software at some point during your 3D printing experience. We have listed some great programs below, many of which are free!

Fusion 360, Tinkercad, Solidworks, Blender, Meshmixer, AutoCAD, Inventor, SketchUp, 123D Apps, 3D Coat, Moment of Inspiration

Figure 2-3. *Chart detailing the hardware and software used in the consumer 3D printing field*

There are many quality repositories that offer free or paid 3D models that can be downloaded, customized, and printed (no 3D modeling experience required!). We've listed a few here to get you started.

SketchFab, GrabCAD, CGTrader, Thingiverse, TurboSquid, 3DLT, Pinshape, Shapeways, 3D Content Central, 3D Warehouse

You don't need a 3D printer to 3D print! There are many companies that offer professional printing services as well as design services to help you bridge the gap from idea to physical form. This is a great option if you wish to have more materials and size options at your disposal.

Sculpteo, Shapeways, ProtoLabs, 3DHubs, HoneyPoint3D, 3D Systems, Studio FATHOM, RedEye

If you need a 3D model of an already existing object, 3D scanning is a very effective method. You can create scans using a simple digital camera, or there is a wide variety of professional solutions available.

Artec3D, MakerBot, Fuel3D, Faro, David-3D, Scansite, 123D Catch, ReCap, HoneyPoint3D

Figure 2-4. *Chart detailing places to find models online, where to get those 3D files printed, and ways to get physical objects back into a 3D file via scanning*

Media & Education

There are various educational resources that teach applicable skills for students and professionals who want to learn more about 3D printing. Furthermore, there are countless books, newsletters, magazines, and forums that will help you stay informed of new trends and developments in the industry. We've listed some popular sources below.

Makezine.com, Make Books, 3DPI.com, 3DPrint.com, 3Ders.org, TCT Magazine, Print Shift, HoneyPoint3D

Finished Products

There are some companies that sell consumer-ready 3D printed products that you can purchase directly. This approach follows the traditional shopping experience where customizability is limited but the design is high quality.

Robohand, Amazon, Shapeways, Nervous System, 3DLT

Retail Shops

3D printing retail shops are a relatively new concept, but are becoming more common as the technology advances. They offer a great physical location where you can buy 3D printers or 3D printing services.

iGo3D, iMakr, The UPS Store, The 3D Printing Store

Figure 2-5. *Chart detailing consumer-focused channels for people to discover more about the 3D printing field*

Industry Event Organizers

Events are a great way to stay informed on the latest developments in the industry. They also offer an opportunity to see 3D printing technologies in person, and in most cases meet the people developing them. These events happen all throughout the year, so check out these organizers to find the next one happening near you!

Media Bistro, Rising Media, Maker Media, IET, 3D Printshow, TCT Magazine, The 3D Printing Association, 3D Print Expo, World Technology Expo, FabCon 3.D, IDTECHEX 3D Printing, LinkedIn, Meetup, RAPID

Community Maker Spaces

If you're looking for a place to go tinker, then a Fablab or hackerspace is for you. Although they carry other maker equipment like CNC machines and laser cutters, you'll likely find a 3D printer that you can experiment with. They offer friendly, community-minded environments for learning.

Tech Shop, The Fab Lab, Fabcafes, Fab Lab San Diego, MakersFactory, Deezmaker, a Community Makerspace near you!

Legal & Market Research

3D printing technology continues to develop. As such, there are many legal and economic issues that come along with it. Here are some companies that are actively researching 3D printing and its worldwide impact.

Wohler's Report, Smartech, Finnegan, Gartner, 3DPI.com, America Makes, 3D Hubs, SENVOL

Figure 2-6. *Organizations helping the 3D printing field grow and reach new users (all infographics by Jeff Hansen, Honey-Point3D™)*

3/How 3D Printing Is Being Used Today

Makers are not the only ones using this technology to further advancements in every area of life. Companies and other organizations have adopted elements of 3D printing to enhance, improve, and even create their products and services. Nothing makes a new technology more real than when you see it being used in everyday life.

In this chapter we'll look briefly at some of the innovative ways individuals and organizations are currently offering 3D printing technology and why it's gaining more momentum.

Rapid Prototyping for Your Ideas, Designs, and Inventions

Have you ever wanted to drive a car that had a steering wheel made just for you? Or how about door handles that were made to resemble the favorite rowing oar you used in college? You can have it made with 3D printing. In fact, Figure 3-1 shows that you can even customize a *whole car* with 3D printing.

Maybe you don't need a 3D printed car, but what about a personalized 3D printed luggage tag or a replacement doorknob? Rapid prototyping is defined as the ability to quickly fabricate a model of a physical part using three-dimensional computer aided design (CAD) software. 3D printing is ideal for rapid prototyping in that you can make changes quickly and produce a sample of one, saving time and money.

Figure 3-1. *Detroit, January 2012, Local Motors Strati, a 3D printed concept car using rapid prototyping (Steve Lagreca, Shutterstock.com)*

In the rapid prototyping division of our company, we see hundreds of people just like you looking at ways to invent and personalize the objects around them. They often come to us with hand drawings, a physical prototype they made by hand, or even just ideas! We then translate this information into a *CAD model* (i.e., a digital file). The first 3D print lets our clients test the design. If adjustments are needed, we modify the CAD file and then 3D print the next version. This process continues until the client is happy. Figure 3-2 shows an example of how one of our clients turned an idea into a phenomenally successful promotional giveaway she gives to her website's membership. Individuals, small businesses, and large corporations everywhere can now customize products in small production runs.

Do you have an invention? Think of your possibilities! Instead of expensive tooling costs that can run into the tens of thousands of dollars, you can now test your ideas with a 3D printer and, in many cases, get your first prototype for under $10 USD per 3D print. And because changing an aspect of the design in a CAD modeling program is easy, each and every object can be subtly different for each buyer. This is something injection molding either cannot do or would require extensive postprocessing in order to achieve.

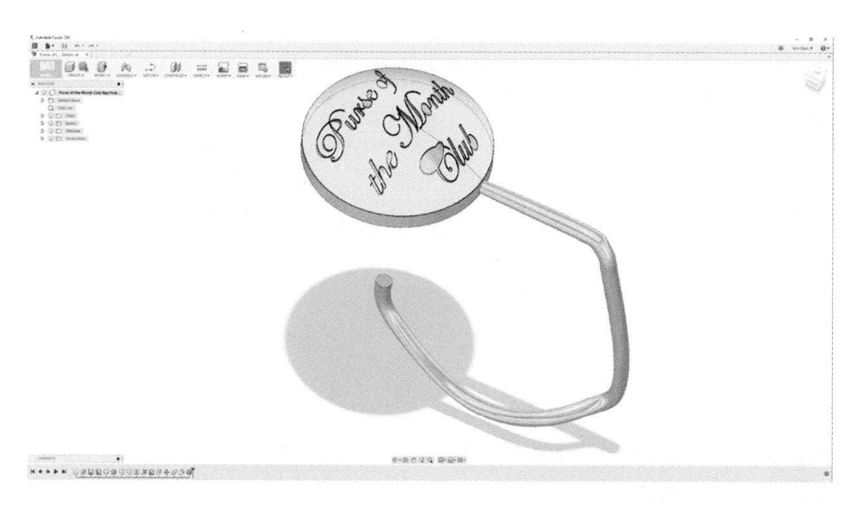

Figure 3-2. *Online retailer Purse Of The Month Club (http://www.purseofthemonthclub.com) created a 3D printed handbag holder as a promotional giveaway to members*

Never before has there been such a versatile manufacturing tool that is so accessible to the masses. Your rapid prototype could be an iteration of a design or the final product. It's up to you!

Forward-Thinking Companies Are Actively Bringing You 3D Printing

As of this writing, companies are offering ways to let customers participate in 3D printing. For example, Lowe's Innovation Labs, a subsidiary of the hardware store giant Lowe's, launched a customer 3D printing initiative in its Mountain View, California, Orchard Supply Hardware store in 2015. This program allows you to work with a Lowe's specialist and design your own hardware items, such as the knob shown in Figure 3-3. Lowe's hardware stores are now providing a venue for you to create with them. You can start with an existing design or make one from scratch, making a version of something into existence that was not there before. It's *your* something, *your* version, *your* twist.

Figure 3-3. *A 3D printing and 3D scanning kiosk in an Orchard Supply Hardware store (image credit: Lowe's)*

In 2013, The UPS Store, the world's largest franchisor of retail shipping, postal, printing, and business service centers with over 4,800 locations in the United States and Canada, started offering local in-store 3D printing services using the Stratasys uPrint SE Plus 3D printer. Now more than 62 UPS Store locations across the country offer up to "same day" 3D printing services to thousands of their customers. The retail giant is expected to expand its 3D printing capabilities in 2016 (Figure 3-4). Overnight, The UPS Store brought local 3D printing to a national level.

Figure 3-4. *The UPS Store currently has 62 locations throughout the United States that offer 3D printing services (photo credit: The UPS Store)*

Don't Worry. It Won't Cost You More. Complexity Is Free

Will companies charge you more for complexity? When you buy something, it's almost always the case that the more complex or ornate versions cost more. This is not true with 3D printing. You don't need to pay more for the highly complicated, complex designs. The 3D printer just "reads" where to put the material... it doesn't "care" if it's simple or complex. The 3D printer just registers how much material is being used, not how intricate the design is.

This is another reason why 3D printing is important. Companies and you can now design highly complex objects (Figure 3-5) that were once impossible to make with traditional manufacturing techniques such as injection molding.

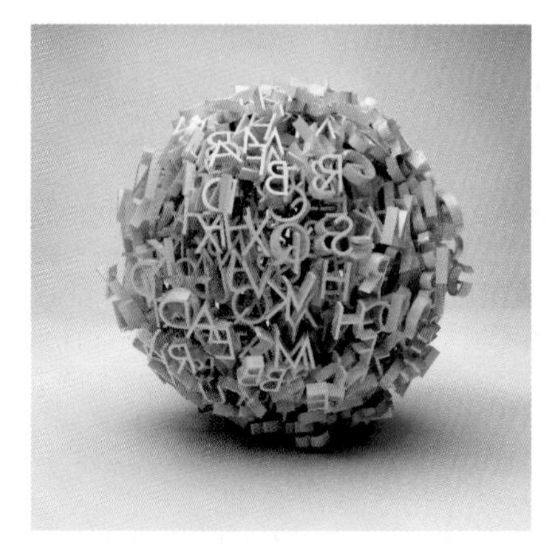

Figure 3-5. *A highly complex 3D printed ball that would be impossible to produce using common manufacturing techniques such as injection molding*

The ball shown in Figure 3-6 has one other smaller ball inside of it, all 3D printed in one step. The ball inside actually moves and is independent of the outer layer!

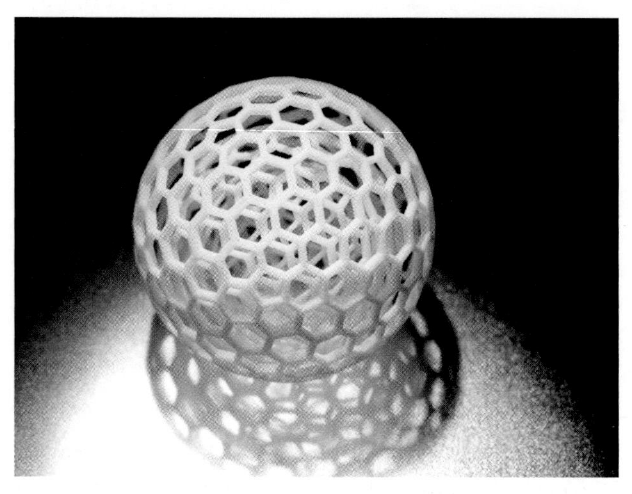

Figure 3-6. *3D printed ball that demonstrates high-complexity manufacturing, normally more costly or impossible to make*

3D Printing Is Advancing Health Care

Have you heard of a company called Invisalign? It manufactures invisible dental braces that are perfectly formed to the shape of your own teeth. Each plastic tray is custom and no two are alike (see Figure 3-7). Dentists make 3D printed molds of your teeth as your alignment changes (hopefully for the better) and then this company produces the braces that are placed on your teeth. Invisalign makes more than 17 million sets of braces a year. New medical applications using 3D printing are being developed by other companies for hip replacements, prosthetics, and more! Not only can you use 3D printing in your life, but it can be a *part* of you, literally!

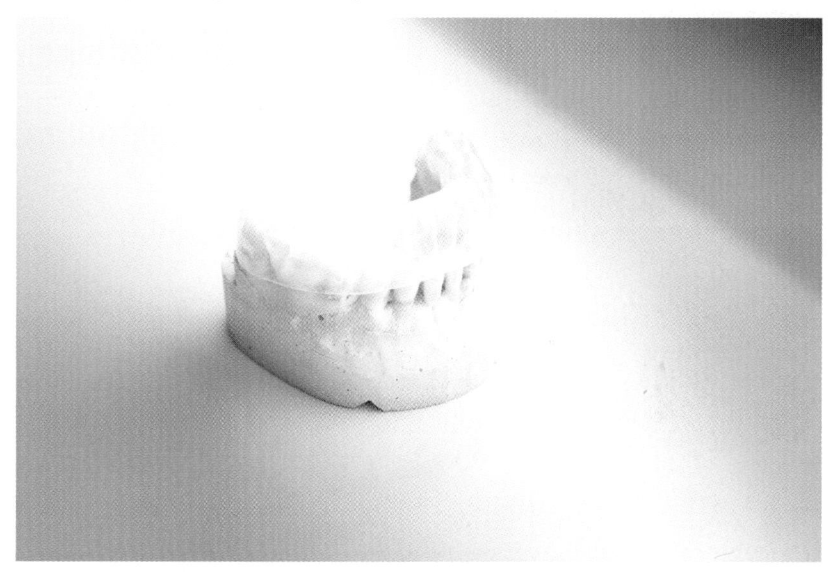

Figure 3-7. *Dentists use 3D printed molds of teeth and jawlines in order to make invisible plastic teeth aligners for patients*

3D Printing Has a Sole

You know the sore feeling your feet have after walking around all day. Some people have even more special needs for their feet due to medical conditions. The answer to alleviating that pain is often custom orthotics, but those are commonly expensive and take some time to produce. A company in Europe called SUNfeet wants to help reduce the amount of a fatigue both your feet and wallet have by providing 3D printed insoles at a fraction of the cost and length of time to produce. Rather than a lengthy fitting and hand fabrication process to receive custom orthotics, a person just has to stand in a 3D foot scanning booth and a 3D model of the foot is made instantly. The 3D model is then used to fabricate a custom insole that is lighter, cheaper, and manufactured more quickly than traditional techniques (Figure 3-8).

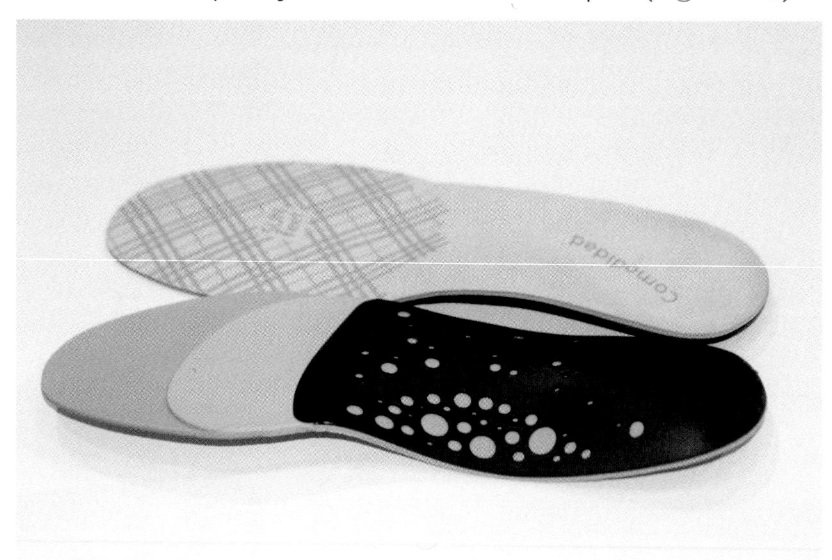

Figure 3-8. *A custom-fitted 3D printed orthotic. The holes are for aesthetics as well as weight reduction, and make those areas slightly more flexible, as dictated by the needs of the patient (photo credit: SUNfeet.es).*

3D printing is ushering in an era of *personal mass customization*. From custom orthotics to custom orthopedics, this technology can create hundreds of thousands of unique items for customers with all of the complex work being taken care of through the software. And the manufacturing is done on-demand, right when you order it! The capabilities demonstrated by these examples illustrate why 3D printing is becoming increasingly relevant in your day-to-day life.

In the next chapter, we'll dive into the hands-on part of this book by describing the most prevalent consumer 3D printing technology: fused deposition modeling, or FDM for short. Turn the page to learn more!

PART II

Hardware and Printing Choices

4/Understanding FDM Printers

There are two main types of consumer 3D printers: those that use filament for fused deposition modeling (called FDM) and those that use resin for stereolithography (called SLA). In this chapter, we will describe FDM technology, which is used in the most common consumer-level 3D printers. The next chapter will talk about SLA printers.

Consumer 3D printers only became readily available around 2009, with much of the growth starting in 2012. Because this industry is so new, you can still see the "roots" of where it originated. As we mentioned in the preface, many (if not all) of the consumer printers being used today came from a community-driven project called the RepRap Project.

RepRap stands for "Replicating Rapid Prototyper." You can learn more about it at RepRap.org (*http://www.reprap.org*). This "Project" is a global community of tech-savvy people that came together to work on 3D printer hardware and software. It seeks to create and refine freely available 3D printer designs and to democratize user access to this world-changing technology. They donated their time and work to the world, and to you.

The first RepRap Project focused on the FDM printer. Many of the current components and manufacturing techniques came from the original work of the RepRap Project. While 3D printer kits that you assemble yourself are still available, you can now buy fully assembled units that are designed to be usable right out of the box. The heritage of the RepRap kit tradition is alive and well, though. If you do decide to build your own printer, you can do so by sourcing all of the parts yourself, or you can purchase a "kit" where you follow assembly instructions.

Any FDM printer will have the components shown in Figure 4-1:

- Filament
- Extruder or extruder assembly
- Build plate / build area
- Linear movement components
- Frame / chassis
- Controller unit

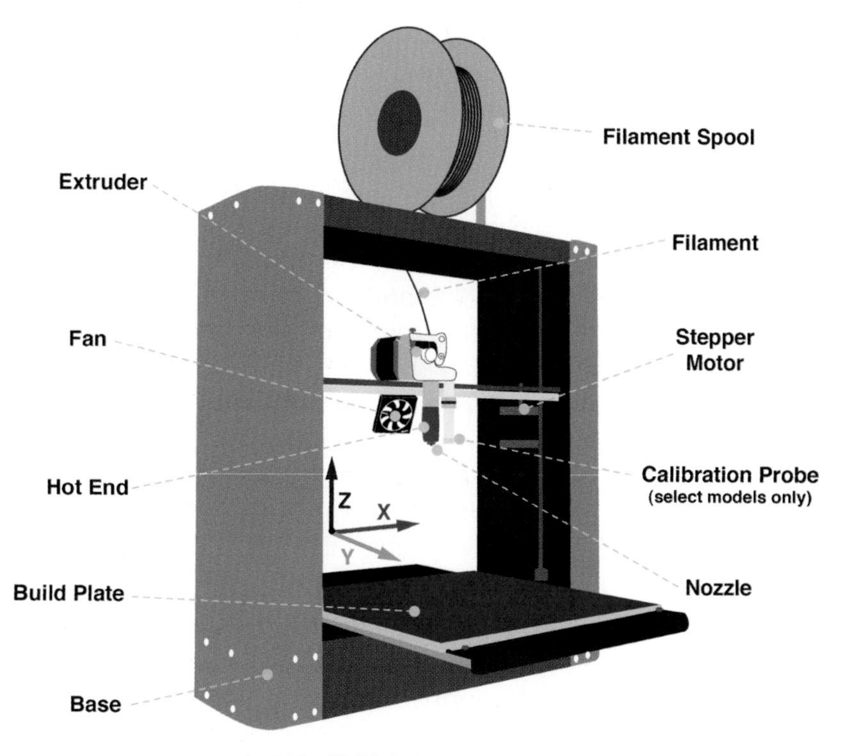

Figure 4-1. *Anatomy of an FDM 3D printer modeled after a Printrbot (infographic by Jeff Hansen, HoneyPoint3D™)*

Here's a basic description of what the parts of an FDM printer do:

- The *build plate* (or *print bed*) is a level, flat area where the 3D print starts.
- A *filament spool* holds the plastic that will be melted.
- Thin *filament*, usually a type of thermoplastic, is wrapped around a filament spool and is the raw material the 3D printer uses to make objects. Consumer printers print with filament that is one of two diameters: 1.75mm or 2.85mm. Diameters are not interchangeable, so be sure to research what your printer uses.
- The *extruder* is the name for the assembly that grabs the filament, and pushes it through a heating block.
- *Stepper motors* drive the belts and extruder up/down, left/right, and forward/back.
- The *chassis* is the frame of the 3D printer, which could be made from metal, plywood, etc.
- The *hot end* is the part of the extruder that heats the filament to just the right temperature based on that material, with a nozzle at the bottom to allow the molten filament to flow through.
- A *fan* controls the temperature at which each layer cools, making the depositing and bonding of the layers more successful.

Take note: There are many consumer 3D printers on the market, with a variety of features and functions. For instance, some have print beds that do all the moving, while others rely on the extruder assembly to do all the movement. All are valid designs and particular choices come down to personal preference.

Resolution Levels

Let's take a moment to talk about how 3D print quality is defined. In the normal 2D printing realm you will see printers that claim "600 dpi" or "1200 dpi" resolution. Those metrics refer to "dots per inch" of ink deposited on the paper. The more little dots of ink per inch, the higher the resolution. It's similar in 3D printing!

All 3D printers (FDM and SLA) have two different types of resolutions:

- *Layer Z* is the height resolution (in the up/down, or Z plane)
- *XY* is the positional resolution (in the left-right, forward-back, XY planes)

The Z-height resolution is the most common metric you will see in FDM printers. As you can see in Figure 4-2, these three cubes are all the same size, but they are 3D printed at different layer resolutions. All of the measurements given here are in "microns," which are fractions of a millimeter. 1,000 microns equal 1 millimeter, so we are talking about resolutions that are typically around 1/5th of a millimeter (or 200 microns).

Figure 4-2. *A close-up image of three layer heights printed at (left to right) 100 microns, 200 microns, and 300 microns*

In the three cubes in the figure, the individual layers are more densely packed on the left than on the middle or right. If you want a higher-quality surface finish, then you will choose a layer height that is smaller, which packs more layers into your end object. The lower the micron level, the less you can see the print lines on the object.

Resolution Versus Speed

An object that prints with a layer height of 150 microns will take twice as long to print as the same object with 300 micron layers. That is because the printer has to print double the number of layers for the same object. The lower the microns, the longer to print, but the higher the print quality.

The other metric for resolution is the XY accuracy—how accurate each individual layer (if viewed by itself from the top) can be "drawn" by the printer. Because most FDM printers use a .4mm nozzle (400 microns) they are limited to features on the edges of each layer of around 250 microns, as shown in Figure 4-3. This is very good resolution, but as you will see in Chapter 5, there are much more accurate technologies available if you need extreme detail.

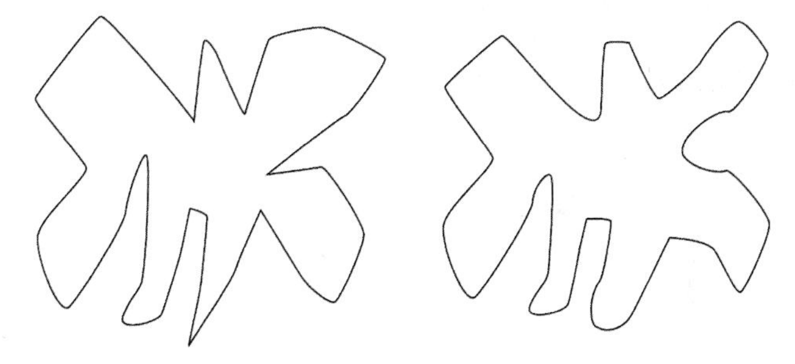

Figure 4-3. *XY positional accuracy between layers created using SLA (left) and FDM (right). Notice less detail on the right image based on thickness of plastic extruded.*

FDM printers do some phenomenal work, though! Look at some examples in Figure 4-4 to see the quality level achieved on FDM 3D printers.

Figure 4-4. *Bracket with no gaps, and perfect layer finish (left). 3D model of "equation of time" based on work from LongNow.org (right). With the 3D print being so detailed, the low resolution of the original 3D model can be seen (right).*

Frame/Chassis

This is what you first see when you look at a 3D printer; the overall general shape. There have been many 3D printers that have enjoyed crowdfunding success on websites such as Kickstarter or Indiegogo because of beautiful 3D printer enclosures/frames. If the enclosure looks nice, people are more inclined to infer it's a quality printer. This is only somewhat true, and this is where your own research can really help you make an informed purchasing choice.

Don't judge a printer by it's cover. Many 3D printers have 3D printed parts holding them together. And why not? It helps keep costs down and allows for easier upgrades later on. While 3D printed parts might lack the durability of machined parts, having 3D printed parts in your printer should not be a mark against the overall printer itself. If an update to the design becomes available, you can print your own upgrades! Keep in mind, 3D printed parts can be found in all price ranges of printers, and some have no 3D printed parts at all. As with all products, it is advisable to read the forums and unbiased reviews to see if a printer, no matter how it is assembled, is reliable.

The frame/chassis needs to be structurally sound over time. 3D printers have moving parts. They get bumped and prodded when you're removing prints from the build plate. If you will be moving your printer from location to location, frame strength is something you will especially want to keep in mind and a metal frame holds better than wood.

There are quite a few "kit" 3D printers that you have to put together yourself that are incredibly structurally sound, such as the Prusa i3 kit and Maker's Tool Works MendelMax 3.0. When you are looking at FDM printers (or really any 3D printer) look critically at the choices that the manufacturers made:

- How thick is the sheet metal forming the chassis?
- Is there flex to the horizontal arm that carries the extruder assembly that would result in potential alignment issues?
- How rigidly is the build plate held in place? How easy is it to loosen and adjust?
- What sort of components went into the extruder assembly? Are they machined parts? Are they custom made, or something that is shared by many other manufacturers?

A good example of a well-made 3D printer is from a San Francisco Bay Area company named Type A Machines (TAM). TAM started its business selling 3D printers that were assembled out of plywood. After its initial printer gained traction, the company wanted to appeal to both hobbyists and low-end manufacturers so it changed the chassis to prebuilt metal. Because TAM decided to move to a 3D printing assembly that had no 3D printed parts, the overall price rose, but the reliability and overall customer satisfaction also rose. As we said earlier, it is not really about the specific components that go into a 3D printer, but how well it works, and how well it holds up over time.

In addition to an all-metal frame, these 3D printers show some good features to look for like custom extruders and ribbon cables (Figure 4-5), as well as silent stepper motor drivers that regulate the electricity sent to the motors to allow for extremely quiet printer operation (Figure 4-6).

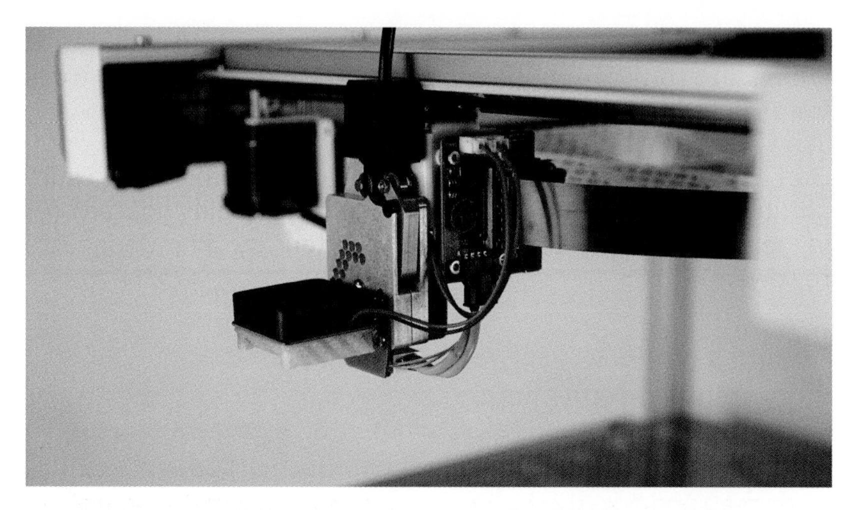

Figure 4-5. *Custom extruder on a Type A Machines printer*

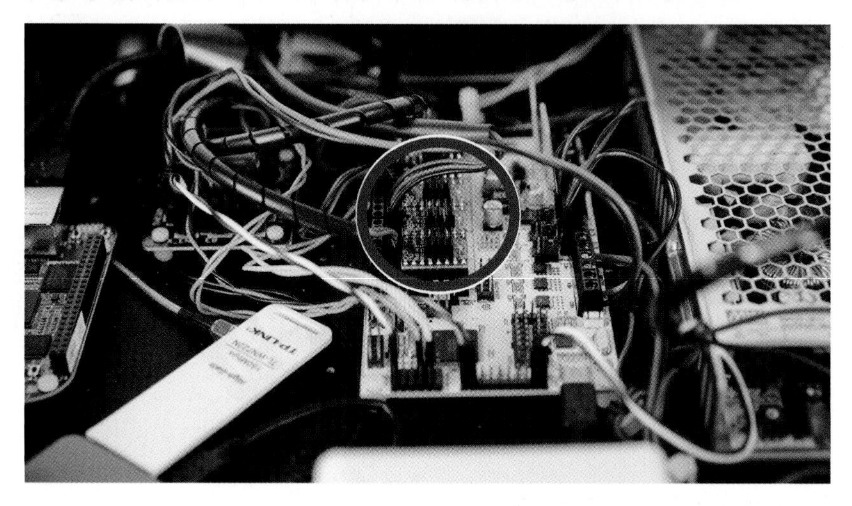

Figure 4-6. *Electronics bay of the Type A Machines printer with silent stepper motor drivers highlighted*

The two figures illustrate choices made by the 3D printer manufacturer to adapt the printer for longevity and printing in office settings. The extruder is one of its own designs, and is rated for higher temperatures than most of the materials you will normally find.

Remember, over 90% of companies in the consumer 3D printing realm are what would be considered "small businesses" with 40 or fewer employees, so the fit and finish of each product is something that characterizes the company behind the product. Don't be fooled by fancy renderings or pictures; there is no substitute for seeing and experiencing the build quality yourself (or in reading independent reviews written by knowledgeable reviewers).

Build Plate

The build plate (or print bed), as shown in Figure 4-7, is pretty straightforward. It is the place where the 3D model is built, layer by layer. Different manufacturers use different materials for the build plate, and you will commonly see acrylic slabs as well as glass. This is an important part of the print process, and one that you should evaluate carefully before you purchase an FDM printer.

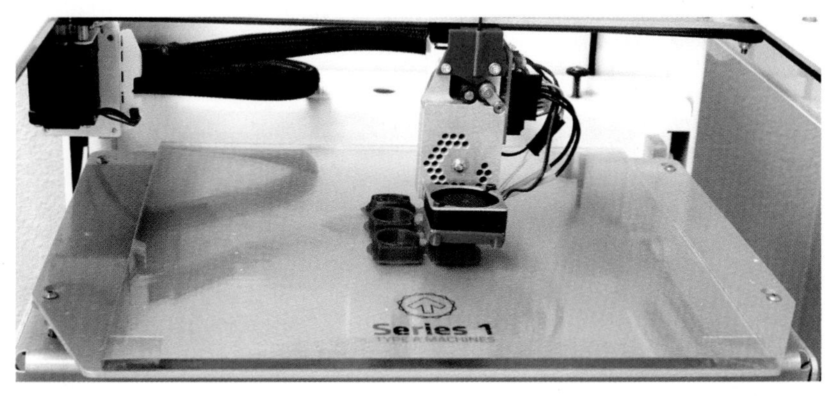

Figure 4-7. *Glass build plate*

All FDM printers require the build plate to be level to the extruder assembly's movement plane(s), otherwise your prints will print lopsided or, worse, fail completely.

Many people think that the term "leveling the build plate" means that the plate needs to be level to the ground. This is not the case. The build plate needs to be level to the travel of the extruder assembly for the print itself to be level. In other words, the extruder and plate need to be perfectly parallel.

Ease of Calibration

When evaluating 3D printers, it is important to research how easy it is to calibrate/level the build plate to the extruder assembly. Today, only a very few FDM printers use automated systems to level the build plate. For the most part, you as the end user will have to level the build plate manually quite often, or risk failed prints.

It is also good to understand that build plates themselves are different. Some are heated, which allow for more materials to be printed on them, and some are nonheated. A picture of a heated build plate is shown in Figure 4-8.

Figure 4-8. *A glass print bed/plate with a red heating element underneath. The heat bed temperature sensor (which monitors the print bed temperature) is to the far right, under the glass.*

You can print a wide variety of materials without a heated bed. Of course, having the *option* to use a heated plate opens up your material selection to more materials but usually comes at an added cost to the base printer price.

You must prepare your build plate in order to get the 3D print to stick to it. Running with a heated build plate sometimes mitigates this, but for the most part you will be using one of a few different methods to get your print to stick. Some people use blue painter's tape, while others use a glue stick; there are many ways to get the print to adhere to the build plate. A more comprehensive description of these techniques is described in "Tips for Success with FDM Printers" on page 49.

Linear Movement Controls

All 3D printers need to use motion to print. There are two general types of 3D printing movement systems, both achieving good build quality but going about the process in very different ways. The two types are called Cartesian and Delta printers.

Cartesian movement systems (Figure 4-9) were the first to appear in the consumer realm, and their movement is based on principles you probably learned in school. There are three axes in a Cartesian printer—X, Y, and Z:

- X movement is to the left and right.
- Y movement is to the front and back.
- Z movement is up and down.

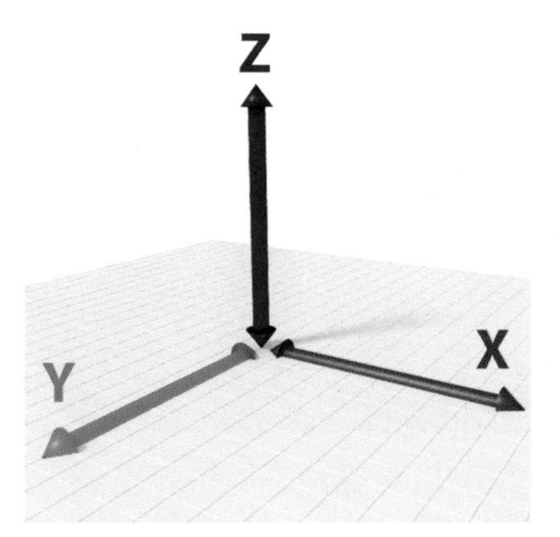

Figure 4-9. *Visual representation of Cartesian coordinates with X, Y, and Z axes labeled (infographic by Jeff Hansen, HoneyPoint3D™)*

These terms can change between printers, though. Some printers will move the nozzle in what seems to you to be a left-right movement, but call that plane of movement the Y plane. Suffice to say that all Cartesian printers use the X/Y/Z coordinate system, drawing out each layer in the XY space, and then moving the extruder assembly by a certain number of microns in the Z direction to start the XY print process over again.

Delta printers go about 3D printing in a different way. These printers borrow from mechanisms that have long been used on on assembly lines called "pick and place" machines. Instead of an X/Y/Z linear system, they use a system based on a "floating" extruder assembly that is pulled in a combination of directions by three arms that are attached to linear rails and pulleys, as shown in Figure 4-10.

Figure 4-10. *The Rostock Max v2 Delta printer from SeeMeCNC. Picture taken looking downward from the top of the printer to the build plate.*

To understand how a Delta printer works, imagine you and two friends holding a hockey puck with a pen in the center of the puck. If you all coordinate your movements, you can draw pictures with that hockey puck, but you all have to work together to get the hockey puck and pen in the right place at the right time.

That is how Delta printers work. If you are having a difficult time visualizing the movement, do an Internet search for Delta printers and watch some videos.

Why were Delta printers created if the Cartesian system was developed first and worked well? Delta printers overcome two major problems the Cartesian systems have. First, Cartesian 3D printers are limited in Z height (the overall vertical height of the print) to the maximum length of the large lead screw that raises or lowers the build plate, or extruder assembly as shown in Figure 4-11.

Figure 4-11. *Photo highlighting one of two lead screws in the Mendel Max 2.0 printer*

Lead screws are expensive, and as they grow in length they become more and more variable in how straight they are. Delta printers don't need to worry about the height of prints because the central extruder assembly glides up on smooth rails pulled by pulleys. Long and smooth linear rails are widely available, so Delta printers are almost always capable of far taller prints than their equivalently priced Cartesian printers.

Second, Delta-style printers often enjoy higher printing speeds than Cartesian printers because, generally, they try to eliminate weight from the extruder assembly as much as possible. While Cartesian printers usually have all of the filament pushing and heating mechanisms directly above the extruder, Delta machines use what is called a Bowden-style extruder to drive the filament. In a Bowden setup, the heated block and nozzle are close to the build plate, but the apparatus that grabs and pushes the filament is located somewhere else, off of the central area. This allows for the movements of Delta printers to be faster because they do not have to push all that weight around.

The drawback to Delta printers is twofold. (These might not be drawbacks to you, so judge accordingly.) Delta printers print on a circular build plate because of how the three arms that pull the extruder move. Accuracy at the edges of the circular build plate tends to drop because one or more of the arms is being pulled to its maximum length, which introduce abnormalities in the print dimensions. Many Delta printers avoid this issue by advising you to only print in a smaller center area of the build plate, or by enlarging the size of the entire Delta printer, and then making the outer areas "off limits" to actual printing. So, while offering greatly enhanced Z height printing capabilities, Delta printers may have a smaller X/Y print area.

The second drawback concerns the use of the Bowden extruder itself. The filament is grabbed by a pinchwheel and pushed through a guide tube (usually made of Teflon) from farther away, as shown in Figure 4-12. This is fine for most filament, but the amount of arc/curvature in the tube can cause problems when, for example, using flexible filaments. This is because the flex in those filaments gets increased over the long distance from the extruder to the hot end, and the filament can press through the heating block unevenly, causing irregular extrusions.

Figure 4-12. *Bowden-style extruder mounted on the back of an Ultimaker 2 printer (http://www.ultimaker.com)*

Extruder

The filament feeds through an extruder (or extruder assembly). The extruder is a collection of parts that feed the filament into a channel, heat up the filament to its melting point, and then force that molten material through a small-diameter nozzle onto the build plate.

The extruder assembly is made up of several parts, as you can see in Figure 4-13. They are described here:

- Drive gear (used for gripping the filament)
- Extruder drive motor (turns the drive gear)
- Filament channel (filament is guided through here)
- Idler bearing (pushes the filament against the drive gear)
- Heating block (melts the filament)
- Nozzle (through which the molten filament flows onto the build plate)
- Cooling fan

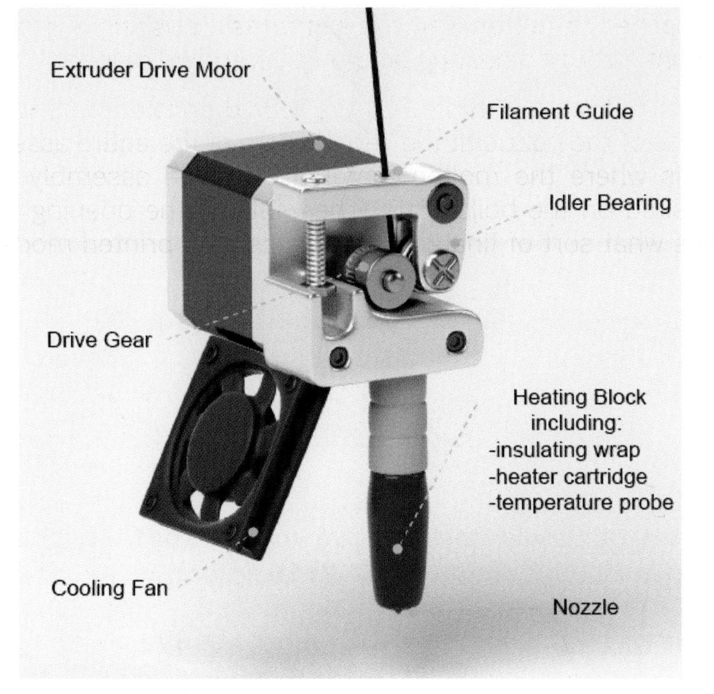

Figure 4-13. *Labeled diagram of extruder components (infographic by Jeff Hansen, HoneyPoint3D™)*

All FDM printers have variations on this setup. Some have extruder assemblies that are single units where all of the complexity is hidden inside one replaceable unit, while other 3D printers choose to go a more modular path to allow users to upgrade and change out pieces as their needs change or upgrades become available.

The following images demonstrate the different types of extruder assemblies different manufacturers use. Figure 4-14 is an extruder from a $3,595 machine (Type A Machines), which is made completely out of machined metal parts. Figure 4-15 is from a $1,500 self-assembled machine (Mendel Max 3) and features a combination of machined and 3D printed parts, and Figure 4-16 is from a $600 pre-assembled machine also with a combination of machined and 3D printed parts (Printrbot Simple Metal). The following three images show slightly different approaches to performing the same task: pushing a strand of filament through a heating block in a controlled manner.

Generally, the only thing that you would consider changing is the size of the nozzle at the very bottom of the entire assembly. This is where the molten material exits the assembly to be deposited on the build plate. The width of the opening determines what sort of final characteristics your printed model will have.

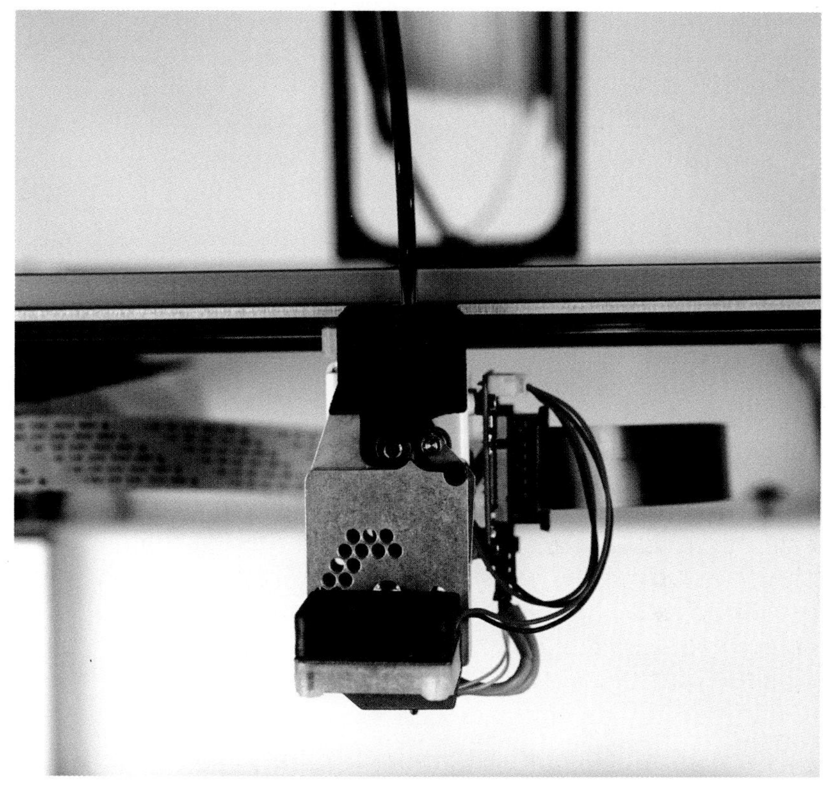

Figure 4-14. *All machined metal G2 Extruder from Type A Machines on the Series 1 printer*

Figure 4-15. *Maxstruder from Maker's Tool Works on the Mendel Max 3.0 printer*

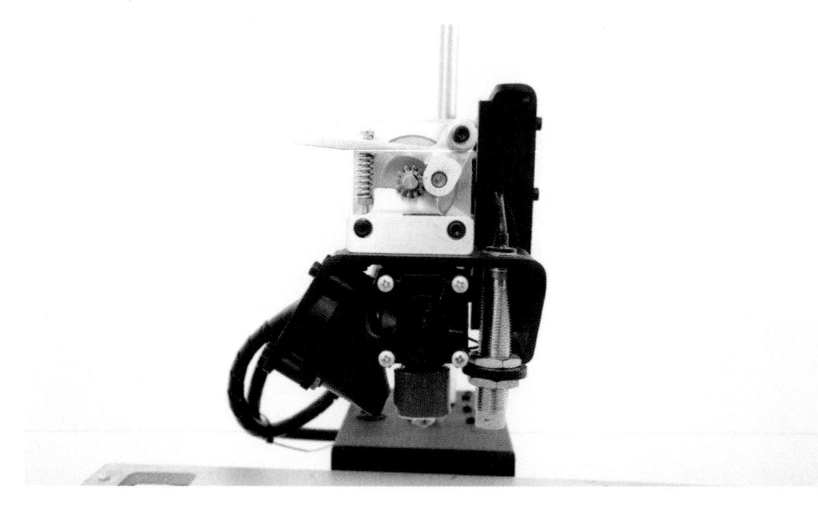

Figure 4-16. *Gearhead extruder on multiple Printrbot models (Printrbot Simple Metal, shown)*

Standard nozzle diameters vary, but are generally around .4mm. These dimensions dictate the width of the standard "line" of filament that comes out from the nozzle, called the *extrusion width*. Strictly speaking, the extrusion width can be set in your printer preferences, but the nozzle size dictates a few important things:

- A finer nozzle allows for thinner walls to be created and for surface features to be more defined (like using a smaller/sharper pencil for fine-detail drawings).
- Physics determines the maximum amount of material that can flow through a specific physical nozzle opening. If you want to print large objects quickly, then you will need a larger nozzle size.

Figure 4-17 gives an example of the size of the individual extruded lines of material. A .4mm nozzle usually creates a single extrusion line that is about .48mm wide.

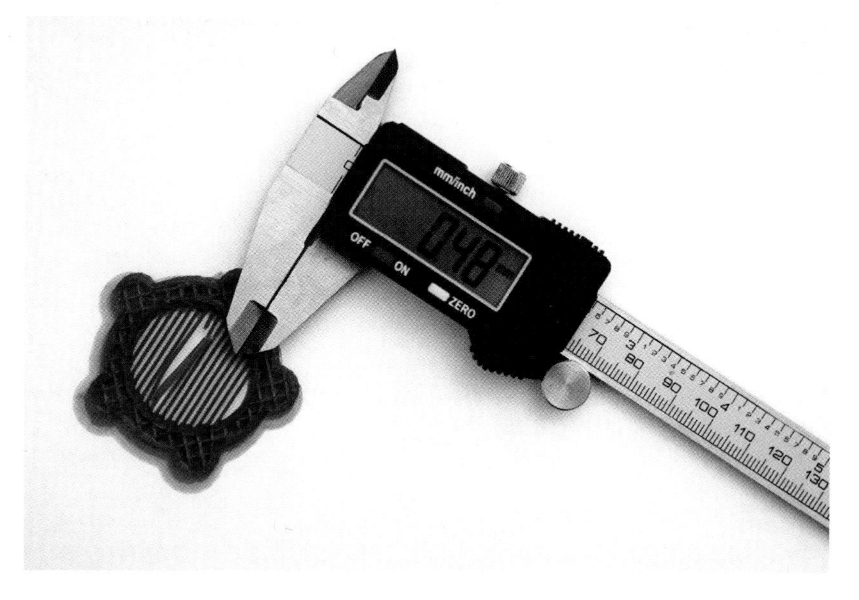

Figure 4-17. *Calipers show an extrusion width of .48mm extruded from a .4mm nozzle*

Any nozzle from .35mm to .45mm is a standard size that is both fast enough to print large objects and small enough to print small features.

If you need to print objects that are more specialized (either larger or smaller), many printers offer a way of changing out the nozzle size to accomplish certain goals. Nozzle sizes of .25mm are pretty slow at printing larger objects, but if you need very thin walls for your prints, you may want to move to a smaller nozzle size like this.

On the other hand, there are nozzle and extruder assemblies such as the one from E3D Online called "The Volcano," which allows for much larger extrusion widths and heights, as you can see in Figure 4-18.

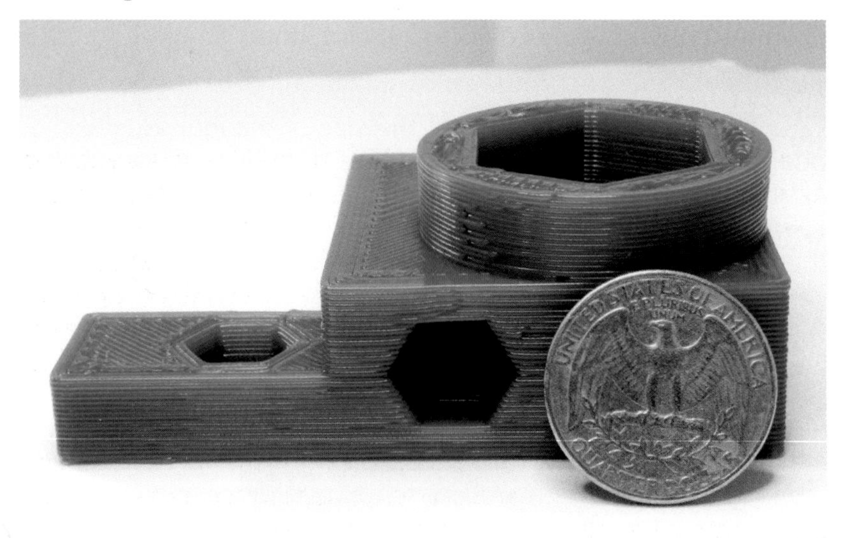

Figure 4-18. *Object printed by E3D Volcano extruder, with .8mm high layers. This part printed in half the time of normal 3D printers equipped with .4mm nozzles.*

The nozzle diameter that printed the object in Figure 4-18 was 1mm wide, and nozzles like that can print very large objects much more quickly than smaller nozzle sizes ever could. If you want the flexibility of picking and choosing what types of nozzles you print with, you will want to consider choosing a 3D printer platform that is open to you switching out nozzle sizes without too much hassle, or without voiding your printer warranty.

Many companies offer one warranty for the printer itself, and another, shorter warranty on the extruder assembly because of the faster wear and tear potential. Typically, both the nozzles and the drive gear (the toothed gear that grabs the filament and pushes it through) are considered "consumables" because they experience the most wear of all 3D printer components. The rate at which you need to replace those parts, however, could vary between once per year to once every several years, depending on how you use your printer and how well you maintain it. If you are going to be 3D printing a lot on an FDM printer, it is a good idea to see how easy it is to swap out or clean those replaceable components.

Filament

One of the core components of an FDM printer is the thin strand of filament that the models are eventually made out of. As we mentioned before, filament used for 3D printing can be seen as similar to the "ink" used in a normal 2D printers. Without filament, an FDM printer would have nothing from which to print. Filament comes in two standard sizes, which describe the diameter of the filament: 1.75mm and 2.85mm (sometimes sold as 3mm in diameter). Your 3D printer will support only one diameter of filament, and you cannot easily use a different size filament if your printer was not designed to use it. Fortunately, there is usually no appreciable price difference between filament diameters, though 1.75mm is a more common filament diameter size. From a technical standpoint, 1.75mm is more common because it requires about six times less force to extrude than the thicker strand. Thinner strands require less bulky, expensive motors to drive the filament through the extruder, which lessens the overall weight of the print head.

There is an ever-expanding range of different materials that can be fed through consumer 3D printers, with the FDM printers offering the most selection out of all printing technologies discussed in this book. The most common material is called polylactic acid (PLA) and is actually a type of nonedible carbohydrate (sugar) derived predominantly from corn. This material prints beautifully and is the starter material of choice for all consumer 3D printers for many reasons.

Several common filament materials are shown in Figure 4-19, but note that some of these materials may or may not work with your 3D printer. It is best to check with your 3D printer manufacturer to see which materials they support.

PLA *Polylactic Acid*

A type of sugar, biodegradable, very stiff but if stressed too much can snap, gets "moldable" at temperatures found in a sealed car in direct sunlight on a hot day. Many different blends of PLA available from glow in the dark to 70% metal content.

TPE *Thermoplastic Elastomer*

Flexible material you can crush by hand and the object will return to initial shape.

ABS *Acrylonitrile Butadiene Styrene*

Same material as what Lego bricks are made of. Durable, and takes stresses well. Smells like burning Styrofoam when printing, petroleum based.

Helper Materials

Materials designed to create dissolvable support structures for use with dual-extruder 3D printers. Examples include PVA (polyvinyl alcohol) and HIPS (high-impact polystyrene).

PET *Polyethylene Terephthalate*

What soda bottles are made out of, a good improvement over ABS. Recyclable, no discernable smell.

Nylon

Extremely durable and good for applications requiring parts to rub against each other as well as for tensile strength and medical applications. Usually harder to print with.

Figure 4-19. *Chart of common FDM materials (infographic by Jeff Hansen, HoneyPoint3D™)*

There is a wide variety of filament on the market, ranging in price from $25/kg to $70/kg. That variation in price does not always reflect a better-quality filament. Be prepared to do some research to inform your choice in materials. Here are some suggestions of things to consider:

- Make sure your printer manufacturer has tested the filament that it sells with the printer you purchased. These filaments will usually be slightly more expensive, but you will know that the manufacturer has done the testing for you.

- Shop online sites that have user reviews. If others have had good experiences with filament from a specific manufacturer, then chances are they will print well for you too. (But also beware that false reviews exist.)
- Buying a spool of filament just based on a picture or the lowest price is risky. We recommend that you purchase filament from manufacturers that are willing to reveal their "story" to you. A good choice is Breathe-3DP (*http://www.breathe-3dp.com*), a manufacturer that reveals how it makes and tests its filament. Whoever you buy from, remember that great CAD models and excellent 3D printers can't save a print from poor-quality filament.

Tips for Success with FDM Printers

Proper storage of your filament is key. Success in 3D printing is all about *variable reduction*, which means limiting the number of things that can change from one printing to the next. Between prints you want your filament to remain as fresh as when it was first delivered to you.

Make sure that dust and debris do not accumulate on your filament, because it will then get drawn into your nozzle, clogging it. Storing filament in a clean, temperature-controlled environment is one of the best ways to protect your filament when not in use.

Many FDM materials, like PLA and nylon, are *hygroscopic* materials, which means they absorb moisture from the ambient air. If you live in a humid area, this could be even more of an issue for you. For all the reasons mentioned, it is best to store filament you are not using in an airtight container like the one shown in Figure 4-20.

Figure 4-20. *Spools of filament stored in a mostly airtight container, with desiccant pouches added*

The desiccant canister you see in Figure 4-21 is inexpensive, can be purchased online, and can be "recharged" if it absorbs too much water by drying it in a kitchen oven for a few hours. This type of desiccant is ideal for filament preservation.

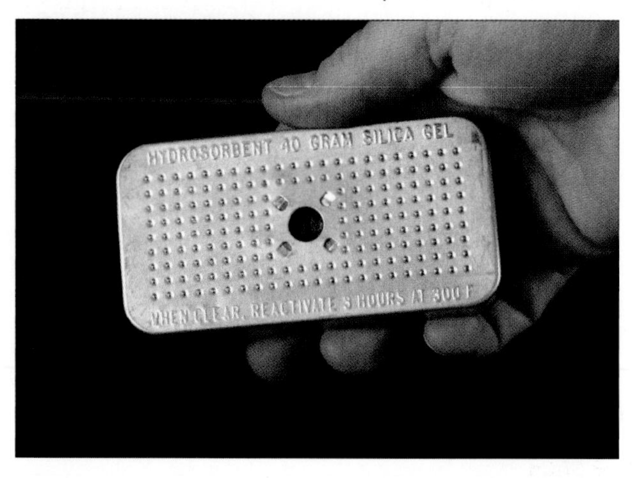

Figure 4-21. *Desiccant canister showing instructions on how to refresh the humidity-absorbing material inside the container*

Build Plate Adhesion

In the 3D printing world, there are many instances where users have found some new and innovative way to make something work, and then that practice spreads to others in the community. Experiencing and finding innovative ways to get 3D prints to stick to the print bed is one such area. Following are some techniques that people have discovered to successfully enhance build plate adhesion:

Blue painter's tape

Applying this tape to the top of the print bed, as shown in Figure 4-22, has been shown to be a good way to get prints to stick. This tape is easily removable, so sometimes the tape can peel up while printing. Unfortunately, it can take your model with it so make sure the edges of the tape are rubbed down well to prevent peeling. If you are lucky, it will not rip off when you remove your print, allowing you to reuse that tape for the next few prints. Blue painter's tape really only works well for PLA prints.

Figure 4-22. *Blue painter's tape applied to the print bed in even, parallel lines*

White glue stick

This is our favorite method to use, as shown in Figure 4-23. This choice is much less expensive than even blue painter's tape. The downside is that it requires you to wash your print bed with a wet paper towel every few prints to remove excess glue buildup. And if your build plate is nonremovable, then you need to be more careful, because you run the risk of dripping water on sensitive electronic components that are underneath. This type of glue is water-soluble, so if you live in a humid climate, or where it rains a lot, it will not perform as well.

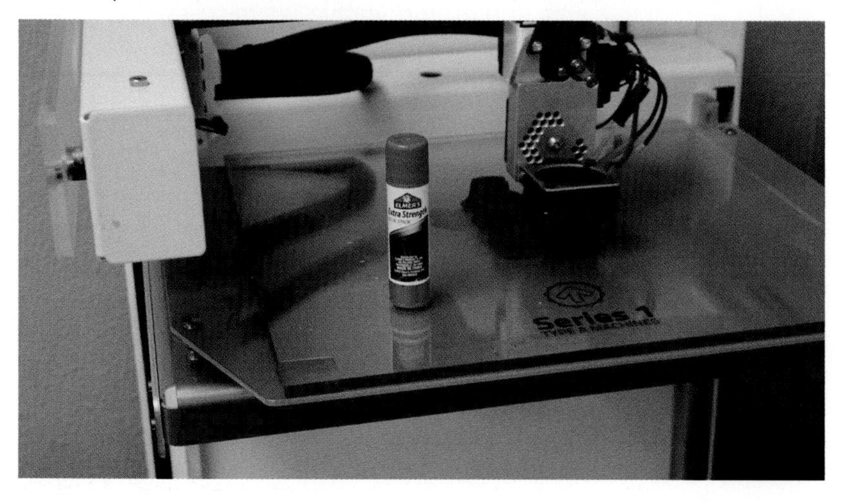

Figure 4-23. *Consumer glue stick used to adhere prints to the build plate*

Kapton tape

This is a transparent, orange, special type of high-temperature tape that was originally designed for use in NASA space missions. It can handle the high temperatures of heated print beds, and is tacky enough for some materials to stick directly to it. Kapton tape is primarily used in conjunction with a heated build plate to allow materials like ABS (and to a lesser extent PLA) to increase print adhesion to the build plate. A thin roll of Kapton tape is shown in Figure 4-24, with many different sizes available to purchase.

Figure 4-24. *Kapton tape*

Specialized 3D printing build surfaces
: These are sold as third-party add-ons that help adhere prints to the build plate, as shown in Figure 4-25. Typically, these surfaces claim that no other adhesive or maintenance is needed other than infrequent cleanings. We have found that they work, but care should be taken to make sure you have proper nozzle alignment to the print bed. Otherwise, you run the risk of the heated nozzle melting one or more grooves into the print surface, necessitating a replacement surface to be ordered.

Figure 4-25. *Specialized Buildtak adhesive surface applied to the build plate*

Slicer Programs

The *slicer* is a type of software that translates the 3D model's geometries into movements the 3D printer can read and perform. It provides various settings like print temperature and speed. The slicer also generates support structures that are designed to keep overhanging parts of your model from drooping. (You'll find a more comprehensive discussion on support structures in Chapter 7.) Most 3D printer manufacturers will suggest a slicer to use, and we recommend that you start with that one. In the instructions, you will find profiles provided for your printer and for the filament that comes from the manufacturer.

Not All Slicers Are Created Equal

As with anything, some slicers are better than others. You may "slice" one model with specific settings in one slicer and then use another slicer with the same settings and get very different end-print qualities.

Following is a list of quality slicers:

- Cura (*http://www.ultimaker.com/en/products/cura-software*): free
- Slic3r (*http://www.slic3r.org*): free
- Repetier Host (*http://www.repetier.com*): free
- KISSlicer (*http://www.kisslicer.com*): free and paid
- Simplify3D (*http://www.simplify3d.com*): paid only
- Autodesk Print Studio (*http://spark.autodesk.com/developers/reference/desktop-applications/print-studio*): free

Unless your printer manufacturer has a proprietary slicer, you have a range of slicers to choose from. If you are willing to spend around $150, a particular favorite of ours is Simplify3D, which slices excellently and is used by hobbyists and professional printers who prefer lots of control over the print process. Figure 4-26 shows tool path (layer pattern) visualizations from Simplify3D and Cura comparing how each would print the same object.

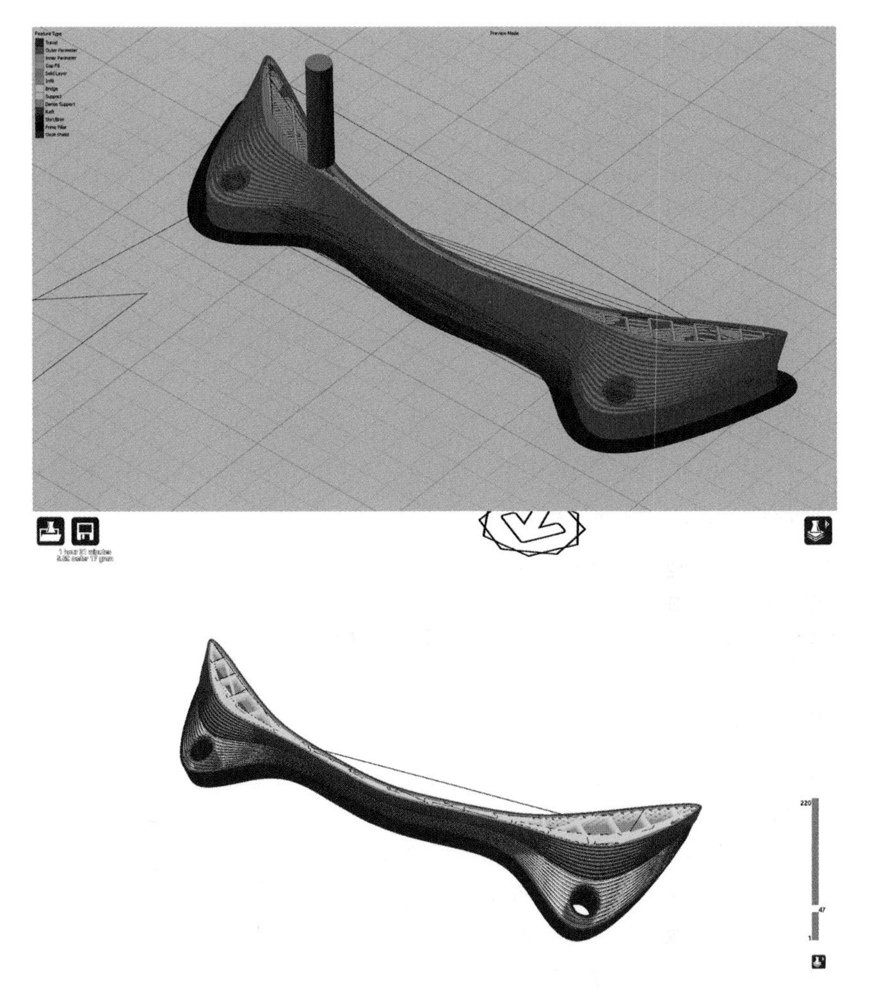

Figure 4-26. *Comparison of tool paths between Simplify3D (top) and Cura (bottom)*

Any of the free slicers will work well for most 3D printers as long as the 3D model you are slicing is a "good" one without errors. While all of them can do some 3D model healing and fixing, you really want to make sure your model is good from the start. (In Chapter 10 we'll discuss how Autodesk's Meshmixer can be used as a tool to help you evaluate 3D models for errors.)

Homing and Leveling the Build Plate

One of the most important factors in encouraging your print to print successfully is making sure that the first layer goes down evenly and "sticks" to the build plate properly. Few FDM printers have automatic bed leveling, so this task will typically be manual.

One of the more popular printers at the time of this writing, the Printrbot Simple Metal, offers what it calls "automatic" bed calibration. There is a sensor next to the nozzle that does *not* set the height of the print bed, but rather reads the angle of the bed and adjusts the print process to compensate for a possibly angled build plate. This prevents one (or more) sides of your print from being shorter than the others. See a picture of the nozzle and sensor from the Printrbot Simple Metal printer in Figure 4-27.

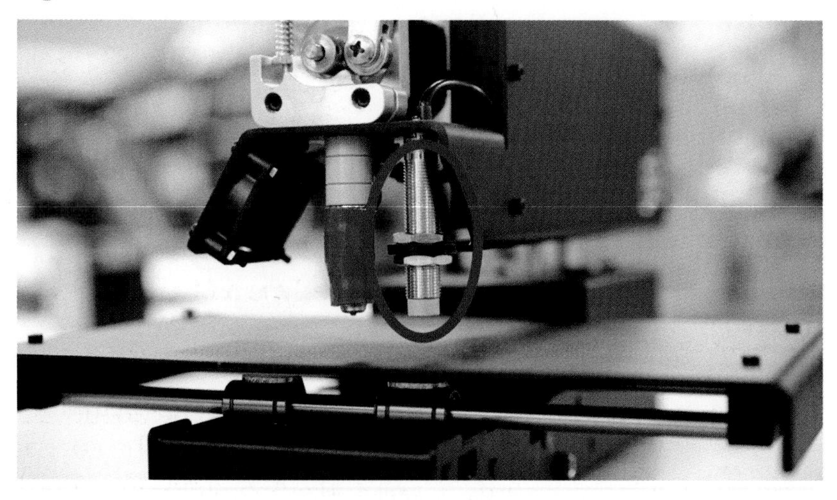

Figure 4-27. *Printrbot Simple Metal with bed sensor highlighted*

Even for this printer, you still have to set the center height of the build plate manually, but the bed itself could be a little off at the edges, and the print will still come out fine.

For other printers you need to do this entire process manually. This involves two steps:

1. Leveling the bed so that all corners create a flat surface in respect to the movement of the left-right (X) and front-back (Y) axes.
2. Adjusting the height of the bed so that the first layer lays down perfectly...not too far away from or too close to the build plate.

Both of the following examples are from the Type A Machines Series 1 printer. Figure 4-28 shows the four screws on each corner underneath the build plate that need to be adjusted to make the build plate level with respect to the movement of the nozzle on the XY axes.

Figure 4-28. *A view from underneath the Type A Machines Series 1 printer, showing the four bed-leveling screws with 3D printed handles*

Once the build plate is level, it (typically) should be about 100 microns, or .1 (one tenth) of a millimeter away from the nozzle, where the nozzle reads it as its "zero" point. A great way to test this is to "home" your nozzle to where it thinks "zero" is, and then try to slip a sheet of paper between the nozzle and print bed. If you can barely slide a normal piece of paper underneath the nozzle, then it's homed. If the nozzle is too close to the plate, the paper will drag significantly under the nozzle or bunch and not pass underneath at all. If the nozzle is too far away the paper will glide underneath with minimal blockage and you will need to make adjustments. See Figure 4-29 for a good "distance" to home your up-down (Z) axis to.

Figure 4-29. *Images showing a piece of paper being slid between the nozzle and build plate with the nozzle being too far away (left) and too close (right)*

Follow Manufacturer Instructions

Our instructions apply to many printers, but yours might be the exception. Always follow your specific printer manufacturer's instructions on how to level the bed and to set the Z height at which you should be printing.

You've now seen how FDM printers work. Driven by the desire for an even better surface finish—one that approaches the quality of injection-molded parts—some users opt for a different kind of 3D printer. We'll examine resin printers and the technology behind them in the next chapter.

5/Understanding SLA Printers

The previous chapter discussed FDM printers, which represent about 90% of the consumer market. The other 10% belongs to a special class of printers called SLA printers. SLA is an acronym for "stereolithographic apparatus," which basically means "gizmo that writes with light." These 3D printers create very detailed prints, but are more difficult to use, so it is important to know their benefits and drawbacks before you buy your first liter of resin.

How It Works

These 3D printers do not use filament; instead, they use a liquid resin (polymer) that hardens or "cures" when exposed to ultraviolet light. The 3D prints from resin printers are still made layer by layer, but are created in a slightly different way.

The basic printing process for an SLA printer is:

1. A vat of liquid resin sits in the 3D printer.
2. A build plate gets submerged to the bottom of the vat, facing downward, until there is a very small layer of resin between the bottom of the vat and the build plate.
3. A controlled light, pointing upward from the bottom of the 3D printer, hits the build plate in a specific pattern, hardening the resin in that specific pattern.
4. The build plate then moves slightly upward.

5. The light shines again in a slightly different pattern from the previous layer, and then cures to the layer that was created before it.
6. The process repeats until the object is complete.

For a graphical representation of the process, see Figure 5-1.

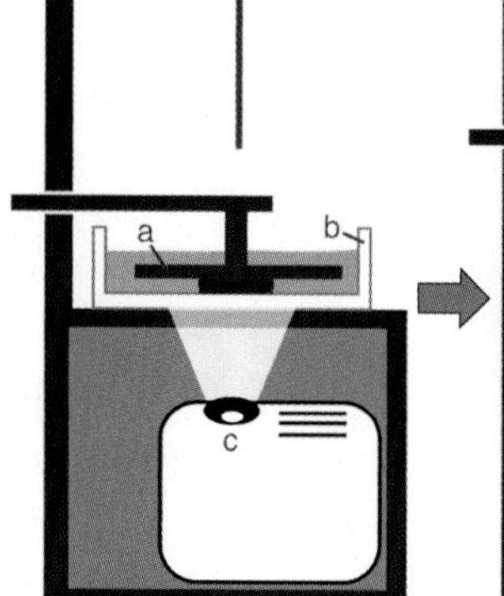

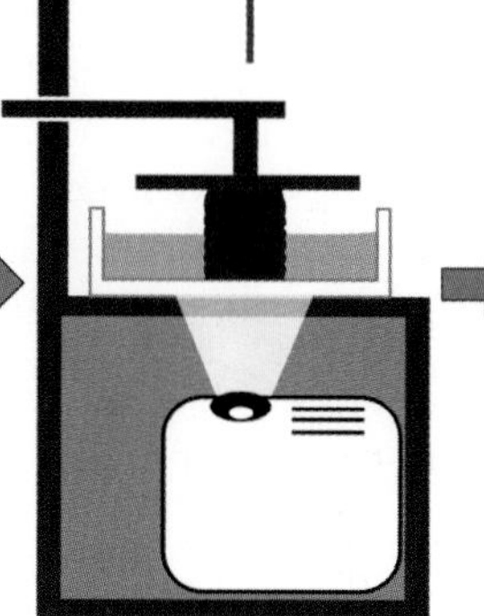

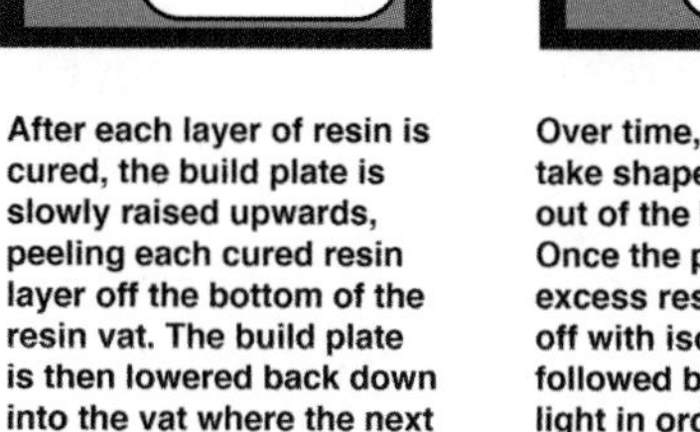

The build plate (a) starts out submerged in a liquid resin vat (b). A laser or a projector (c) shines light through the bottom of the resin vat, which cures the resin in a very precise shape, making it solid.
After each layer of resin is cured, the build plate is slowly raised upwards, peeling each cured resin layer off the bottom of the resin vat. The build plate is then lowered back down into the vat where the next layer will be cured.
Over time, the 3D print will take shape as it is pulled out of the liquid resin vat. Once the print is removed, excess resin must be rinsed off with isopropyl alcohol, followed by exposure to UV light in order to finish the curing process.

Figure 5-1. *The SLA printing process (infographic by Jeff Hansen, HoneyPoint3D™)*

The goal of this chapter is to show the differences between FDM and SLA printers, as well as point out some unique considerations when running an SLA printer. Here are some notable differences between running an SLA printer versus an FDM printer:

- The resin must be kept at around 70 degrees Fahrenheit in order for the viscosity of the resin to not be affected.
- The resin is smelly...a kind of sweet, but strong chemical smell that may bother some people.

- The printer must be kept away from windows because any extra sunlight might cure some of the resin inadvertently.
- You must wear gloves when handling the resin and the new 3D print, as well as for cleaning up the printer to avoid getting resin on your hands. This means, essentially, wearing gloves every time you physically interact with the printer.
- Kids and pets should be kept away from the 3D printer and resin. The resin should be considered as dangerous as household bleach.
- After the print is finished, it will need to be rinsed by hand in isopropyl alcohol.
- Once the print is washed off, it is a good idea to place the print in direct sunlight or in a UV curing box for 10 minutes to finish curing the outer layers.
- Once you are done printing, you will need to remove tiny cured particles of resin that may have appeared and could interfere with future prints. You should pour the leftover resin through a commonly available paint filter or into another container and label that container "used resin" (not mixing it back in with the unused resin). Resin keeps for a long time, but it is good to use up the resin that has already been exposed to light before using brand new resin.
- Unless you plan on storing the resin in the vat (which may get dust in it or spill if knocked over), you should wipe the resin vat clean with a small nonabrasive wiping tool.
- You will need to replace the vat over time. The vat that holds the resin is considered a consumable, with a typical life span of about 3–4 liters of resin. Vats range in price but average around $50–$80 each.

As you can see, there are a lot of considerations you need to take into account with a resin printer. If you are willing to put up with the process, however, you can enjoy truly exceptional prints, as shown in Figure 5-2 and Figure 5-3.

Figure 5-2. *Eiffel Tower model printed on an SLA printer (3D model credit: Pranav Panchal)*

Figure 5-3. *Close-up photo of Eiffel Tower with each individual railing measuring just .2mm*

Cost of Materials

Resin is also relatively more expensive than FDM filaments, though they are slowly coming down in price. You can expect to pay widely variable rates for resin. The least expensive resin can go for $50 per liter (1000 mL) and the most expensive general-purpose resin can go for $150 per liter.

As a rough statement, the price of 3D printing with a $50 resin is about on par with printing in PLA on an FDM printer when measured per cubic centimeter. The more expensive resins are similar to printing with specialty FDM filaments (like PET).

If you are considering only the cost of resin, those prices can seem a bit high. But if you think about your new-found ability to print detailed objects at home and on demand, changing those designs as often as you want, and then being able to successfully cast them, you should consider the price of this process as it existed *before* 3D printing. A full description of the costs associated with 3D printing in resin at home can be found in Chapter 12.

An Open Source Resin Printer

In 2015, Autodesk launched its resin printer called Ember. In the months that followed, Autodesk open-sourced every aspect of this printer, including the formula for the resin. This allowed people to make their own resin and resin printers, creating a drop in price. Most people will not become overnight chemists and start mixing their own UV resin, but if you want to do so, you can read about the formulation here (*http://bit.ly/1ptliqH*).

Types of Resins

Because of the way resin is formulated, there is less choice in material properties for the SLA printers than what you will find in FDM printers. The most common types of resins fall into these four categories:

- Normal/general-use
- Hard/durable
- Flexible
- Castable

How Castable Resins Are Used

Most castable resins need to go through a burn-out process. The resin print is placed in an investment-casting material (plaster) that is hardened, and then the mold is placed in a kiln to burn out the resin inside. This leaves a hollow cavity, and the molten metal used for casting is then poured into that. Castable resins are predominantly used in jewelry design and production.

There are almost always trade-offs and benefits when moving from one type of resin to another. For example, resins that are made for strength and durability cannot print in fine details like normal, general-use resins. But, those more durable resins can take more physical stress than the general-use resins

Some 3D printer manufacturers also manufacture their own resin, and they (usually) strongly suggest that only their specific brand of resin is to be used in their printers. Here is a list of popular resin vendors with their website addresses.

- Formlabs Resin (*http://www.formlabs.com*)
- MadeSolid Resin (*http://www.madesolid.com*)
- B9 Resins (*http://www.b9c.com*)
- MakerJuice Resin (*http://www.makerjuice.com*)
- Autodesk Resin (*http://www.ember.autodesk.com*)
- Spot-A Materials (*http://www.spotamaterials.com*)

Third-Party Resins

While most SLA printer manufacturers allow the use of third-party resin, other manufacturers state in their terms and conditions that the printer warranty will be void if third-party resin is used. We recommend you read the terms and conditions of any printer you want to buy thoroughly to make sure you can use third-party resins.

The Two Types of SLA Printers: Laser and DLP

Laser-based SLA printers trace out the curing path of the resin with a laser, while digital light processing (DLP) printers use a light projector to cure one entire layer at a time. The following chart explains the strengths and drawbacks of each type of printer.

LASER SLA Printer	DLP SLA Printer
Very fast for small objects because the laser only has to trace a small area before moving the build plate up.	Similar speed for all 3D prints, regardless of object size, since the entire layer is either cured or not cured all at once.
Slower for large objects because the laser has to trace large areas.	Much faster, comparatively, for large objects because the entire layer is cured all at once.
Universally come with proprietary software for printing because it is difficult to control the laser accurately with third-party software.	Comes either with proprietary software or a limited selection of free/open source software, depending on manufacturer.
Example: Formlabs Form 2 printer.	Example: Kudo3D Titan 1 printer, and Autodesk Ember.

As a general statement, resin printers, per volume of printable area, are more expensive than FDM printers. While some good FDM printers can be bought for $500, resin printers only really start to come to the market as solid machines at around $1,600 and go up to around $3,500. (Prices may change over time.)

Printer Profile: Autodesk Ember (DLP Technology)

In 2015, Autodesk released the Ember resin printer at a cost of around $7,500 USD (for a complete kit with extra resin vats, cleaning supplies, and starter liters of resin). Ember, as shown in Figure 5-4, is a custom-designed DLP printer that has a very small build envelope of just 64mm x 40mm x 134mm. This small build envelope is perfect for small items like jewelry prototypes and biomedical applications.

Figure 5-4. *The Autodesk Ember DLP printer*

Though this printer has a smaller build envelope, bear in mind that traditional professional-level 3D printers catering to jewelry artists can cost around $50,000. With a $7,500 price tag (at the time this was written), this printer offers an attractive alternative to the previous options, providing the very fine level of detail seen in Figure 5-5.

Figure 5-5. *3D Print from Ember printer showcasing extremely detailed hair-like structures (image by Scott Grunewald, 3dprintingindustry.com)*

What is truly unique about the Ember printer is that every component of the printer has been made open to the public. You can visit the Autodesk website to download every schematic of the hardware, the electronics, the software, and the resin to build on your own, all of which is offered under a nonrestrictive license. The full CAD design files for the Ember printer have been released in Fusion 360, which we will introduce you to in Chapter 11. You are free to build your own Ember printer using those designs (Figure 5-6) if you wish.

Compared to the 150+ FDM printers currently on the market, there are fewer SLA printers to choose from. Resin printing in the consumer realm is less developed partly because SLA technologies exited patent protection much later than FDM printers and partly because the printers are more difficult to run and cost more to buy. The availability of resin printers is quickly rising, though, and represents one of the highest growth areas of consumer printers. Again, the level of detail produced in these printers is unparalleled in the consumer FDM printer world.

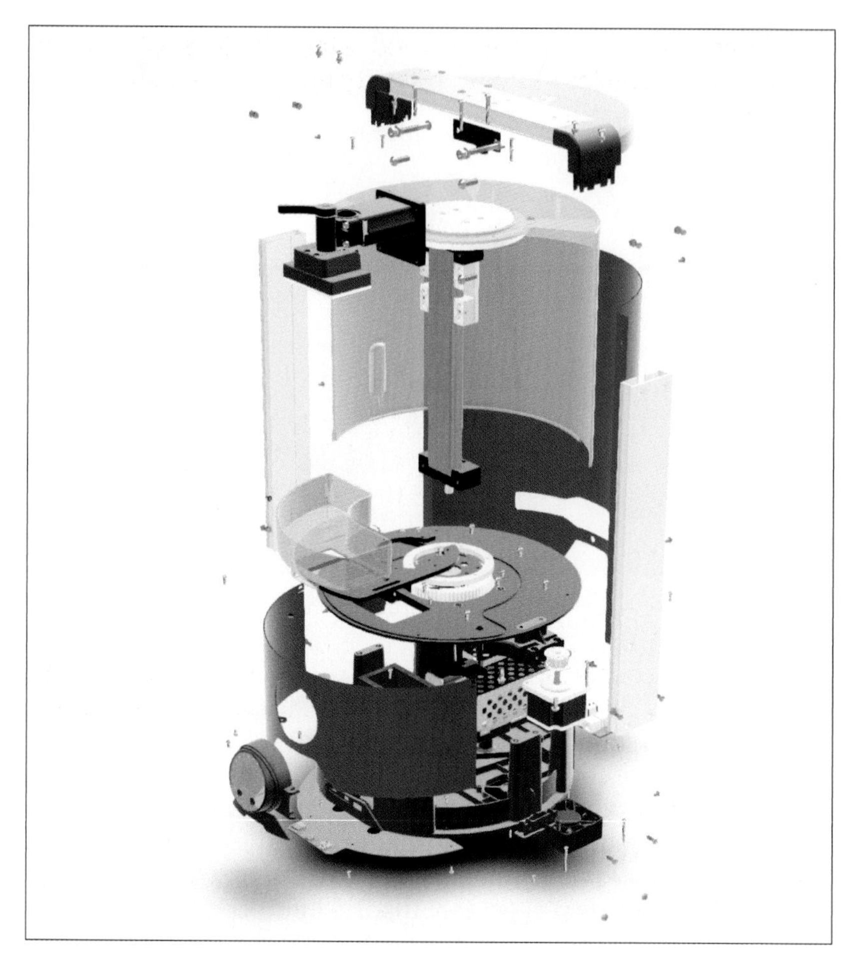

Figure 5-6. *CAD model of Ember printer in "exploded" view as shown in Fusion 360 software program*

Printer Profile: Kudo3D Titan 1 (DLP Technology)

Many 3D printers enjoy success on the crowdfunding website Kickstarter. The Kudo3D Titan 1 resin 3D printer (*http://www.kudo3d.com*) was one such example. Its campaign raised nearly $700K USD to fund the printer's development. The printer had an end price to consumers of around $3,200 USD.

The Titan 1 (Figure 5-7) is a good example of a DLP-based printer that allows for use of third-party resin.

Figure 5-7. *Kudo3D Titan 1 3D printer*

If you look inside the Kudo3D Titan 1 printer, you will see that an upward-facing projector shines its light through a transparent vat onto a build plate, as shown in Figure 5-8.

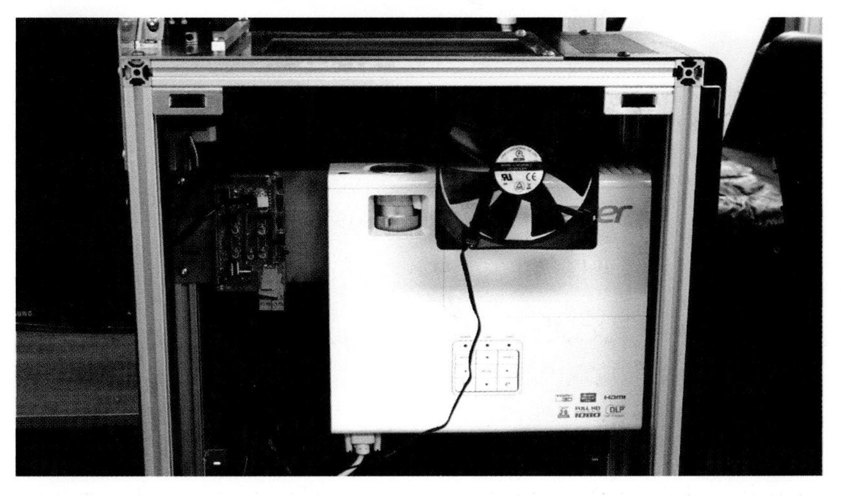

Figure 5-8. *Inside the Kudo3D Titan 1 3D printer with side panel removed, showing the upward-facing light projector*

This printer differs from resin printers like the FormLabs and Ember printers by being more "open" with the software that can be used to generate prints and by using more consumer-available parts (like a modified desktop computer chassis, as seen in Figure 5-8). This allows for larger overall prints. Plus, any slicing software that supports DLP printers can be used, any third-party resin can be used, and the quality from the printer itself is excellent, as shown in Figure 5-9.

Figure 5-9. *Make: magazine's twisted-rook "torture" test, used to push the capabilities of resin printers, printed here with the Kudo3D (http://www.youmagine.com/designs/make-rook-2015-3d-printer-shoot-out-sla-test)*

You will need to calibrate this type of open source printer for different Z-height resolutions by manually moving the projector up or down, and then securing the projector by hand-tightening the screws. Kudo3D claims a super-fast print speed of 2.7 inches per hour at 100 micron resolution for any object up to 7.5" x 4.3" by 10" tall. The fast prints and lower price point make this a great choice for certain users.

Printer Profile: FormLabs Form 2 (Laser Technology)

On the other side of the SLA spectrum is the FormLabs Form 2 printer shown in Figure 5-10. This company also started as a successful Kickstarter and has since brought out multiple iterations of its original laser printer.

Figure 5-10. *FormLabs Form 2 printer*

This printer is used by professionals around the world for rapid prototyping and is an extremely reliable printer. As of this writing it is priced at around $3,500 USD. FormLabs creates its own software for use with this printer and is able to control the print workflow more precisely. In many professional settings, the slightly increased cost of the FormLabs printer, and FormLabs resin, is well received due to the consistent prints that can be relied on for rapid prototyping and end-consumer satisfaction with the build quality.

The laser-based system shown in Figure 5-11 is extremely accurate in detail (see Figure 5-12). While it might not be as fast as

DLP printers when printing large objects, in professional settings it's excellent at creating consistent prints that require a less manual process.

Figure 5-11. *Laser tracing an object on the Form2 printer (photo courtesy John Biggs, TechCrunch)*

Figure 5-12. *Sample prints from a FormLabs Form 2 printer. These objects were printed separately and then hand-assembled.*

Software: Slicers for Resin Printing

Because the printing process using SLA printers can be affected by many factors, many printer manufacturers have created their own software to manage the print process. Typically, that software is tuned to print with a specific formulation of resin, and if you use third-party resin, you may run into issues. This is especially true with all laser-based resin printers, because the software understands how to move the laser in different directions to cure the layers.

Some proprietary software will allow you to tweak settings for exposure time or layer separation time, allowing for other resins to be used, but some software really does not open those capabilities to you. As mentioned before, check with the specific printer manufacturer you are researching to see how well it welcomes experimentation from its printer owners.

Conversely, having an all-in-one software system for resin printing can be a benefit. The FormLabs printers use proprietary software called "PreForm" that functions as a slicer for their resin printers. This software makes the preparation of models straightforward and does a great job at creating support structures for the prints while offering convenient presets for FormLabs resins.

Slicers for DLP printers work a little more simply than having to create pathways for lasers to trace. As you read earlier, these types of 3D printers shine a pattern on the build plate. Technically these patterns are just a series of black and white images. Where the image is white, more light is hitting that area and getting cured into a solid. Any place that is black will not be cured and will remain liquid. In Figure 5-13 you see a typical set of layers that would be projected on the build plate, one at a time, to form an object.

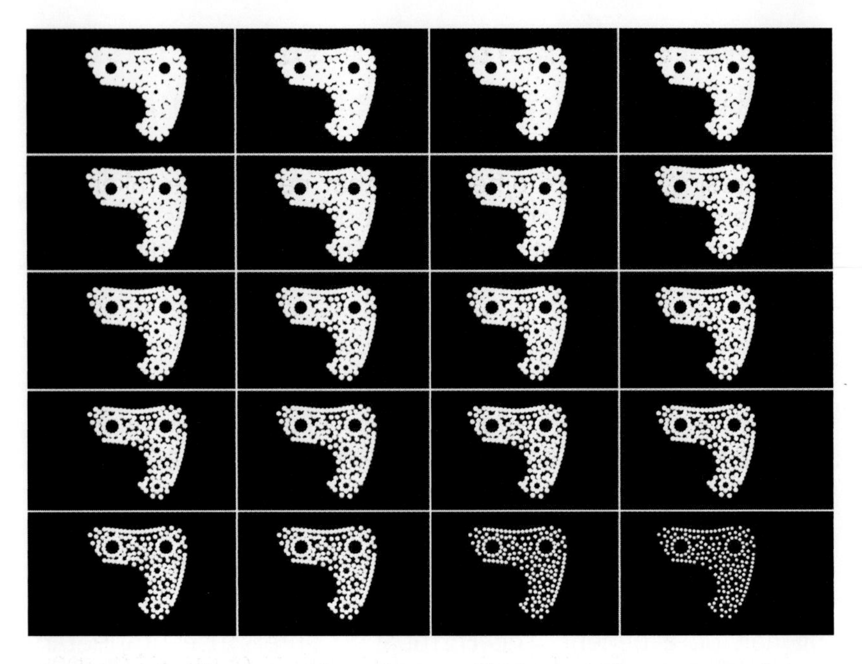

Figure 5-13. *A succession of layers that would be projected, one by one, on the build plate to create a 3D object*

Printers like the Kudo3D, and other "DIY" 3D printers like the LittleRP (*http://www.littlerp.com*); $600 without projector and the DropLit DLP Printer (*http://www.seemecnc.com*); $400 without projector—provide a couple of options for free software, but that wasn't always the case.

For quite some time there was only one choice for free software to create the slices for DLP printers, Creation Workshop (*http://www.envisionlabs.net*): it is able to create support structures, as well as sliced images such as those shown in Figure 5-13.

Now, there is another choice. Late in 2015, Autodesk released its free Print Studio software, which also acts as a slicer. It supports not only DLP printers and their image-slices, but also the tool paths required for FDM printing, all from one central application. You can find it here (*http://www.spark.autodesk.com/developers/reference/desktop-applications/print-studio*).

Support Structures for SLA Printing

Most SLA printers slowly pull an object, upside down, out of a vat of resin. There are differences that need to be taken into consideration for the creation of support structures using this technology versus FDM technology. The three main differences are:

- The model is being printed upside down, so the effects of gravity on your model's overhangs (compared to FDM) will be reversed.
- SLA printers can print much more accurately than FDM printers, and so the support structures tend to be much more delicate and thin.
- All SLA prints experience adhesion between the newly cured layer touching the bottom of the vat and the build plate, or previously printed layers. In between actual printing, the printer will "rip" the print off of the bottom of the vat by movings upward, and then move downward again for the next layer to be cured. All resin prints have a platform that is printed first to prevent those strong forces from being applied to your model. Figure 5-14 shows the difference between SLA and PLA support structures, including the first layer of supports seen in SLA prints.

Conclusion

SLA printers require more discipline in order to get to a final print, but for many the process is worth it. Certainly, for those printing with castable resins in the jewelry industry, the benefits of a resin printer cannot be overstated. Due to the level of detail, SLA prints are immediately more acceptable as finished products for most customers.

You've now learned about FDM and SLA printers. Maybe you are thinking you're not quite ready to take on the challenges of operating and maintaining your own quite yet. The next chapter discusses the pros and cons of owning a 3D printer versus outsourcing the actual 3D printing to a service.

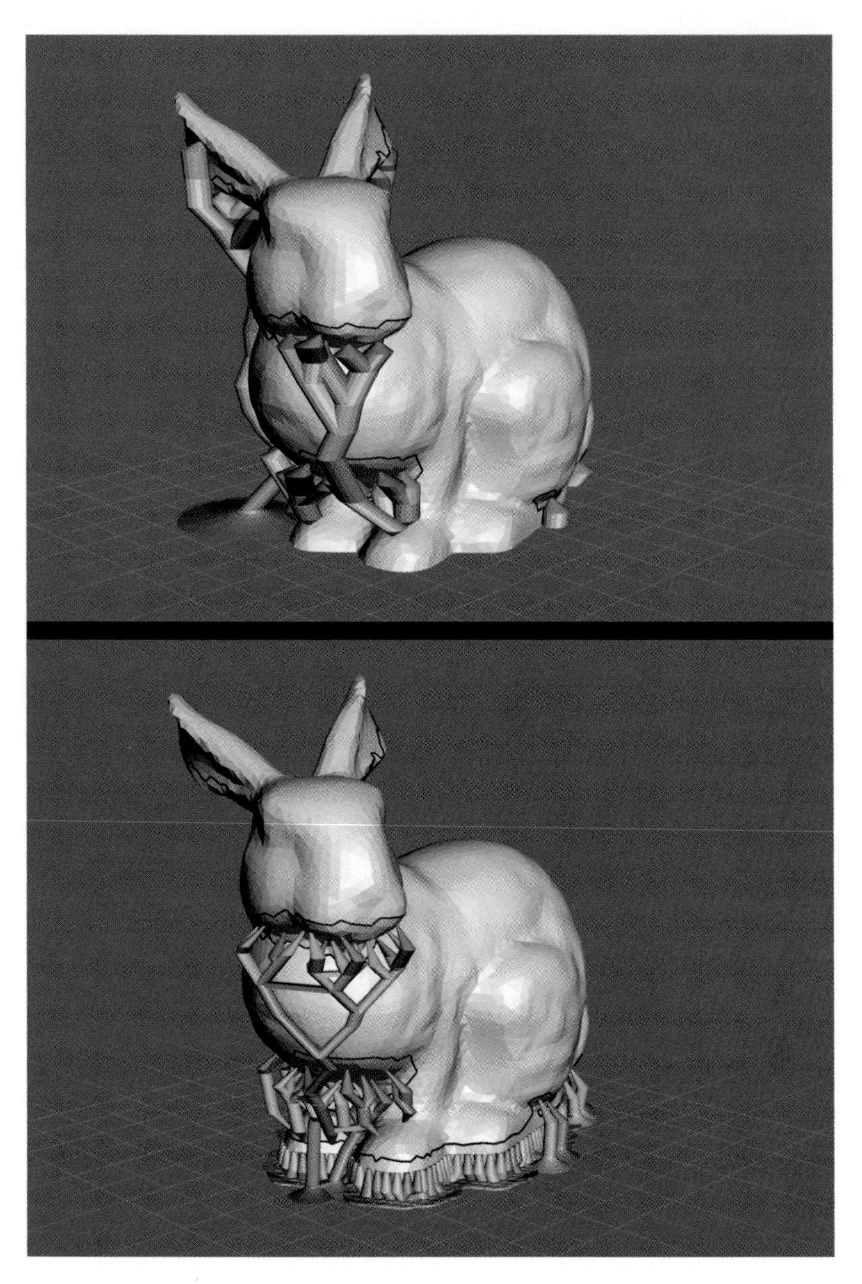

Figure 5-14. *Differences in support structures generated in Meshmixer with the same overhang settings for an FDM printer (top) and SLA printer (bottom)*

6/Outsourcing Versus Buying Your Own 3D Printer

Many people have compared the current state of 3D printing to where personal computers were in the 1980s. Back then, PCs were bigger and clunkier, and the processes were not as streamlined. Hobbyists, who didn't mind fixing the process as they went, were the first to adopt the new technologies. The primary owners of 3D printers today are also DIYers. But that will change. As with personal computers, in time 3D printers will become more of an everyday commodity.

Let's compare the different technologies and options as they are now. We will look at home printing versus outsourced printing and make distinctions between FDM and SLA, as well as consumer versus professional outsourcing (Figure 6-1).

	Home Printing		Outsourced Printing	
	FDM	SLA	Consumer	Professional
Cost	$	$$	$$$	$$$$
Speed Of Delivery	Fastest	Fast	Slow	Slowest
Material Availability	✓✓✓✓	✓	✓✓✓	✓✓
Setup Time			n/a	n/a

Figure 6-1. *The advantages and disadvantages of home versus outsourced 3D printing (infographic by Jeff Hansen, HoneyPoint3D™)*

The Benefits of Outsourcing Your Printing

For many, 3D printing will never happen at home. After all, it is very easy to submit a 3D model online, and have your object arrive on your doorstep 7–10 days later. That object will be printed on a 3D printer that may cost anywhere from $60,000 to $1,000,000.

The graph in Figure 6-1 shows that outsourcing your 3D printing jobs saves you time. You don't have to set up the printer and monitor the print as it's being produced. You also get the benefit of having more material choices. Speed of delivery is less attractive because you have to wait for your print to be mailed to you, and the cost is higher because you are paying for material *and* someone's else's time.

But the highest benefit of outsourcing is that it allows consumers to create objects of the highest levels of quality on some of the most expensive machines in the market. The quality of those prints can be near "production quality," making this not only an attractive approach but necessary for some applications.

Many businesses could benefit from custom-created 3D printed products. Promotional giveaways, rapid prototypes of a new product, creation of visual props for sales calls, and many other examples show the value of 3D printing. But not all these companies have in-house CAD designers and 3D printers. They want to participate in the benefits of 3D printing but can't allocate the financial and time resources to bringing on staff for it. That's where outsourcing really has its benefits.

Luckily, there are many options for outsourcing, and it offers an alternative to otherwise costly productions that would be out of the question for many consumers and businesses.

Popular Outsourced 3D Printing Service Bureaus

Consumer-focused 3D printing services like Shapeways (*http://www.shapeways.com*), shown in Figure 6-2, and Sculpteo (*http://www.sculpteo.com*), shown in Figure 6-3, offer lower costs than the more professional alternatives, and they also offer a wide range of materials and a more accessible ordering process. Businesses are using these companies, too, to make rapid prototypes and small production runs.

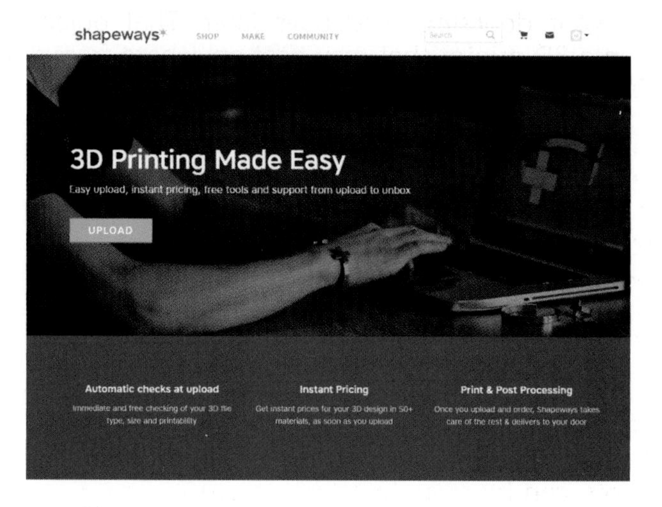

Figure 6-2. *Shapeways home page*

Figure 6-3. *Sculpteo home page*

If you need specialized techniques or absolute precision in your final model, then moving to a more business-oriented outsourced 3D printing service bureau might be a good option. Proto Labs (*http://www.protolabs.com*), shown in Figure 6-4, does not have a repository of designs you can purchase like Shapeways and Sculpteo, but it is focused on more high-end fabrication technologies for customers with specific mechanical and physical properties requirements.

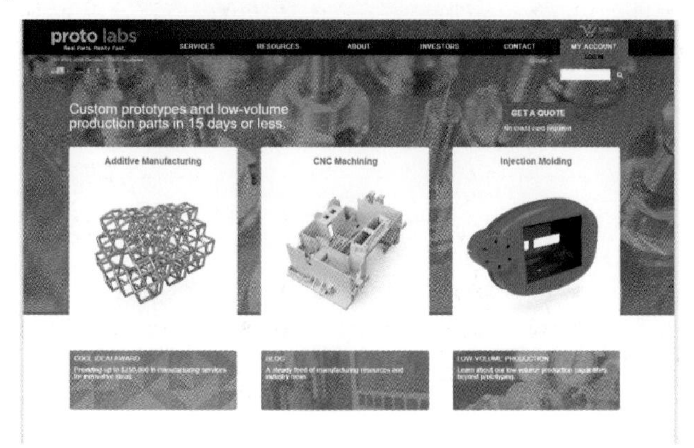

Figure 6-4. *Proto Labs home page*

Proto Labs excels in high-quality steel and alloy 3D printing, CNC machining parts for functional testing, and injection molding for low-volume production runs. If you were a large design firm that wanted to get functional prototypes of new kitchen faucets that needed to screw into existing receptacles, Proto Labs would be the kind of service you would look to. In fact, services like Proto Labs are commissioned to 3D print parts for advanced aerospace designs that will be tested in flight, with large, expensive machines depending on those parts to be strong and stable.

Services like Proto Labs may not be in the public's eye, but they are behind the scenes pushing the medical field into new areas, for example by creating new implants for hip replacement surgeries. They are typically more involved with the high-profile 3D printing stories you might have heard about on the news.

The Rise of Local Outsourcing

Until this point, we have mentioned only large international print service bureaus. Just as with any consumer product, you don't know where your object is being printed, or by whom—you just receive it in the mail. If you want to know who is making your 3D printed object, there is a service called 3D Hubs (*http://www.3dhubs.com*) that delivers locally made prints. Figure 6-5 shows 3D Hubs' home page. Its website is filled with useful information like local events you can attend, industry stats on materials usage, and even a place for you to list your 3D printer as a service!

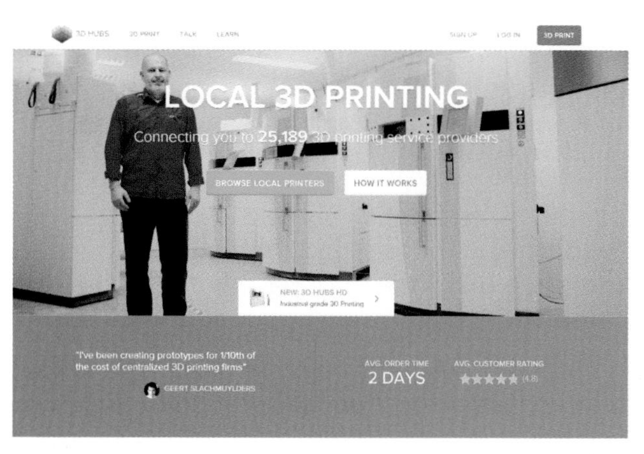

Figure 6-5. *The home page for 3D Hubs showing (at the time of this writing) 25,189 printers ready to make your object locally*

People from around the world have listed their 3D printers on the 3D Hubs platform. Each printer provider, usually just a one-person operation, offers prices and material options. The prices and setup fees vary from "hub" to "hub" so do your research and review their individual ratings before making a purchase.

3D Hubs works similarly to the other outsourcing service bureaus, but with some notable differences:

- You are supporting your local economy by purchasing from 3D printers in your area.

- Most printers on this service are consumer 3D printers (cost under $5,000) so you might be limited in material choices and quality.
- You can pick up your print in a day or two at the "hub" (someone's house), or the "operator" (owner) of the "hub" can ship the print to you.
- You get access to slightly less expensive materials like PLA as well as ones that are more exotic/experimental like hybrid carbon fiber materials, flexible materials, and even, in some areas, chocolate!
- The "hub" or person you order from "lives and survives" on feedback and customer ratings, so they are really motivated to give you good service.
- You get to connect with people in your local area who are knowledgeable about 3D printing and who may have other resources they can refer you to.

Similar to the international service bureaus, 3D Hubs also quotes you an instant price. This might include a flat "setup" fee and the charge to make the print based on the cubic centimeter volume of the 3D model.

After you order, you will receive an email or phone call within a day or so to confirm your order and the printing process starts. Remember that the hub "operators" on this service are local people who own a 3D printer and probably have other day jobs. Because their printer, the filament they use, and their work practices are not regulated by 3D Hubs, we recommend you read the reviews and go with a "hub" that matches your level of expectation. Our experience so far has been good with printers on this service, and the 3D Hubs website is easy to navigate.

If you don't mind searching for a good, self-found printing partner at a cheaper cost, then try 3D Hubs. If you are looking for assurances of material quality, or are working under strict business processes and protecting sensitive designs to comply with your own company's internal processes, then purchasing from the international service bureaus may be a better choice for you.

The Benefits of 3D Printing at Home

Certainly many of the 3D prints you see on the Internet today were produced in people's garages and homes. 3D printing at home offers the hobbyists and makers faster feedback on their 3D model, as shown in Figure 6-6. They can print each version at home, not having to wait days to receive the print as is the case with service bureaus.

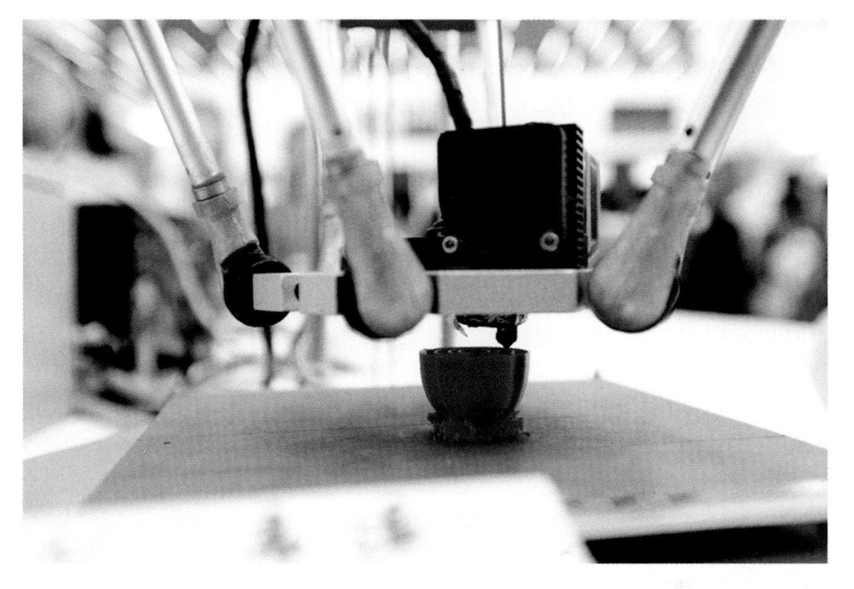

Figure 6-6. *3D printing at home gives hobbyists quicker feedback on a 3D model*

Where 3D printing at home costs around four cents per cubic centimeter, it is not uncommon to see a price of 50 cents to $1 per cubic centimeter from an online service bureau (plus a setup fee added on top of that). That represents a 10x increase in price over printing it yourself, not counting your own time for setting up, monitoring the print, and cleaning up.

A disadvantage for 3D printing at home is one that is also difficult to quantify. It's the amount of time that it takes you to do everything that "surrounds" the actual 3D printing. This includes leveling the build plate, making sure your filament is stored properly, and other factors that take time and some knowledge in order to do successfully. For the most part, once you have gone through your initial learning process for 3D printing at home, the process becomes easier. The analogy is like going to a restaurant versus making the food at home from scratch!

But as we mentioned in the beginning of the book, the act and process of 3D printing at home and being a DIY hobbyist is very rewarding. Remember the last time you learned a new skill—woodworking or making ceramics, for instance—and then mastered it enough to get the results you wanted? 3D printing at home provides you with similar opportunities to expand your knowledge and the satisfaction of honing your skills.

Variables Involved When 3D Printing at Home

There are things to consider if you want to print at home. Let's run down the list of the differences between printing at home with an FDM printer and printing at home with an SLA printer (Figure 6-7).

	FDM	SLA
Printer Cost	$300 to $3500	$1500 to $3500
Printable Size	Large	Medium
Print Quality	Decent	Very High
Print Speed	Acceptable	Slow to very fast
Materials Cost	$20 to $50/kg	$55 to $150/L
Materials Cost per Unit	2.5¢ to 6¢ / cm^3	6¢ to 20¢ / cm^3
Setup/Cleanup	Relatively easy	Relatively difficult

Figure 6-7. *Advantages and disadvantages for FDM printers compared to SLA printers for printing at home. The chart includes the general prices for printers in the consumer realm (infographic by Jeff Hansen, HoneyPoint3D™).*

Price

On average, FDM printers will cost less than SLA (resin) printers. At the time of this writing, a good FDM printer can cost $600 and up, while a good SLA printer averages $2,000. There are some SLA printers that cost only a few hundred dollars, but they require you to provide your own projector.

Size

Some SLA printers that are DLP-based (and not laser-based) will have resolution that slightly decreases when you go to print larger objects. Mainstream FDM printers can print up to 12" x 12" x 12" while many resin printers print comfortably at only a smaller 6" x 6" x 6" size.

Quality

What the SLA printers lack in volume of print area, they more than make up for in the quality of the resulting print. SLA/resin prints from consumer printers are such good quality that they can look injection molded. FDM printers, on the other hand, produce visible layers.

Speed

FDM printers are limited to an upper limit of about 100–150mm/second print speed, having to trace out each line in each layer individually with molten material. Laser-based resin printers work the same way, but they trace with a laser beam and are typically much faster when completing a layer. But one of the drawbacks of resin printing is that the model needs to be separated from the bottom of the resin vat; this process requires force to break away the cured layer of resin and then move the build plate back down for another curing operation to get the next layer. This puts laser-based systems roughly on par with FDM printers.

DLP resin printers (which cure with light) cure an entire layer, small or large, at one time. For larger prints, or prints that have a lot of one thing copied over and over (such as jewelry designs), DLP printers can be faster than laser SLA or FDM printers.

Material availability

FDM printers offer the most diversity with more than 50 materials (and growing) to choose from. 3D printing service bureaus come in second place with the most variety because they own many different types of printers. Some of these services are even willing to 3D print parts in a castable material and then traditionally cast them using materials such as brass, bronze, or steel.

Materials cost per unit

When you create a model, the space inside of the model can be described using a metric of volume, which is typically "cubic centimeters." Online 3D printing service bureaus will give you prices for materials that are charged at a starting flat rate, and then the number of cubic centimeters is multiplied by the price per cubic centimeter.

If you are printing at home, you have no startup fee and your material cost will be much less. A single spool of PLA filament, 1kg (2.2lb), will represent about 800 cubic centimeters of printable material and costs range around $21 to $35. More exotic filaments like PET or flexible filament are roughly double the price of PLA filament, thus in the $70 range.

Resin printing at home costs over 200% more than FDM printing at home based on the three general price points currently offered by resin retailers of $55, or $100, or $150 per liter of resin. Using the same formula for resin's specific density, one liter of resin contains about 900 cubic centimeters of printable material.

Time needed for setup/cleanup

This is the other critical factor that sways people in one direction or the other. With the online print services, the base startup fee for an FDM print can be around $5. For resin prints, startup fees can be around $20. Why the difference? Resin is much harder to work with than FDM. When you run a resin printer, you have to use gloves to prevent the resin from getting on your hands, as well as pour the unused resin back into containers for long-term storage. A resin print also needs to be washed off with isopropyl alcohol to remove the uncured resin, and it needs some time in an ultraviolet curing box (or to be placed in direct sunshine) to finish curing.

For an FDM print, you just pop the print off of the build plate and everything is pretty much done. Of course, if your workflow requires extreme detail, or you do not mind working with resin, you will enjoy the prints that resin printers create.

Production Quantity: Low-Volume Manufacturing

If you are looking to create many units of a design, you have three options: 3D print them at home, go through a service bureau, or use traditional manufacturing. Typically, creating a mold for injection molding (traditional manufacturing) costs anywhere from $2,000 to $4,000 for a simple mold, and then the prices rise for molds that are more complex or that are made for producing many thousands of items. Before the capabilities of 3D printing, there was a gap in the market where inventors and product designers had very few inexpensive options if they wanted to produce a thousand or less copies of an item. The economics for traditional manufacturing (with molds) only made financial sense when production neared the "several thousands" unit range.

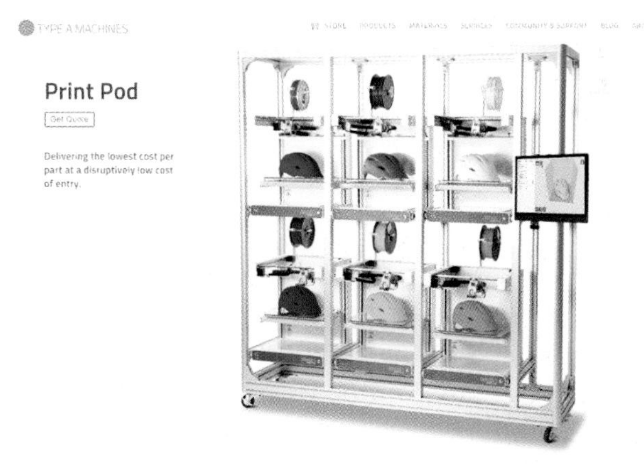

Figure 6-8. *Type A Machines' "Print Pod" offers six printers in a rack with central management for low-volume manufacturing*

Type A Machines not only manufactures 3D printers but also offers production capabilities for runs of 1,000 units or less with the creation of its "Print Pods," as shown in Figure 6-8.

For Beginners, Choose FDM First

For first-time buyers, we recommend starting with FDM over SLA. A good 3D printer in the $400 range is exemplified in the Printrbot Play shown in Figure 6-9. It has a smaller build volume with a 4" x 4" x 5" build envelope (101mm x 101mm x 127mm), but the construction is solid and the price is good.

Figure 6-9. *The Printrbot Play from Printrbot.com*

Printrbot (*http://www.printrbot.com*) is one of the largest consumer 3D printer manufacturers and has made a name for itself by creating durable, quality printers that come pre-assembled at the factory and feature a durable, all-metal chassis.

You can use open source software to create the 3D designs as well as to control the slicing process itself, and the Play has an auto-leveling probe that greatly simplifies the process for build-plate leveling. The Printrbot warranty is 90 days from purchase date for all components. What you learn on this printer will also be applicable to other, more expensive FDM printers, so it's a good choice for beginners.

Choosing the Right Quality Finish for Prints

If you want to sell products to consumers and the quality of surface finish is important, then you have a couple of choices. You can outsource your printing to one of the local or international 3D printing service bureaus to get excellent quality at a higher cost per unit. Or, you can print the objects yourself.

The first option is to use an FDM printer. If you remember from previous chapters, FDM prints show individual layers, and the end consumer's perception of your design may suffer because those layers do not equate to "regular quality." In this instance you would have a couple of options.

One option is to print with a specific material (filament) called ABS (acrylonitrile butadiene styrene). As previously discussed, ABS is a durable material but can be a bit smelly to print with; it also requires a heated print bed. ABS has one unique capability, however. The chemical acetone "liquifies" ABS. You can use acetone vapors to "melt" the outside layers of your print, as shown in Figure 6-10, to create a smooth finish.

Figure 6-10. *Before and after results of treating the outside of an ABS print with a few seconds of acetone vapor*

Acetone: Proceed with Caution

Acetone is an aggressive chemical that is not recommended for inhaling or direct skin contact. The acetone vapor you will be producing by heating the acetone is extremely flammable. Exercise caution, otherwise explosions or fires may occur.

The other method for making production parts out of FDM prints is to do post-print hand-processing to them. This could include sanding the prints to remove the obvious layers created by the printing process, as well as painting the model to cover up the remaining layers. It is a laborious process, but one that can yield nice parts, at the expense of your time.

Evaluating for Special Mechanical Considerations

Even if your product does not need to undergo extremes in terms of durability or performance under stress, something as small as a pair of earrings still undergoes the forces that are common in life. Objects get dropped, or brush up against a wall, or get tossed in buckets with other items. You need to make sure that your print can withstand the environment in which it will be placed. On average, it's best to start with FDM prints, even if those prints might not have the best surface quality as compared to resin prints.

The main reason to prefer FDM over resin is the wide variety of FDM materials available to the consumer, which gives you more choice in suiting the print to its end goal.

Figure 6-11 shows an example of one of the better filament manufacturers, and why FDM printers will be at an advantage over other technologies for mechanical prototyping.

This is a spool of PLA (polylactic acid) from a company named Breathe-3DP (*http://www.breathe-3dp.com*). This specific type of PLA has a "++" in its marketing name to denote that this company feels it has some benefits over "normal" PLA. We have tried it and really seen great results.

Figure 6-11. *A spool of PLA++ from Breathe-3DP*

The filament in the figure is an example of a "blended PLA" that was designed to improve on a characteristic of PLA to be rigid, but brittle when exposed to stresses. There is a growing field of "functional filaments" created by a number of reputable companies that can create not only decorative items, but items that are designed from the 3D modeling stage to take advantage of specific material properties in certain types of FDM-created parts. Some noteworthy brands of "specialist" filament manufacturers are:

- NinjaFlex (*http://www.ninjaflex3d.com*) is flexible.
- PET (*http://www.madesolid.com*) is what soda bottles are made out of and is recyclable.
- PLA++ (*http://www.breathe-3dp.com*) is impact-modified PLA.
- Nylon (*http://www.taulman3d.com*) is very strong and slightly more difficult to print with.

As an example of what a functional filament is designed to do let's take PLA. PLA is a great material to make strong, rigid objects. But, once your object gets stressed to a failure point, normal PLA will snap or shatter. Look at the following example prints to see how a normal PLA part (left, in black) fails while the "enhanced" PLA++ from Breathe-3DP (right, in blue) survives.

Each sample was bent backward and forward three times, as shown in Figure 6-12.

Figure 6-12. *After three back-and-forth bends, normal PLA (left) snaps, while the enhanced Breathe-3DP PLA++ (right) remains structurally sound*

The rise of enhanced/functional materials that overcome the limitations found in current filaments will increase. We will continue to see advancements in both FDM and SLA materials, making them more adaptable and useful in more applications.

Conclusion

You are now more familiar with the options you have when 3D printing your models. You will need to consider your resources in terms of money, time, and willingness to work with technology that sometimes fails when making the decision to print at home or outsource the job to a third party.

It's great fun to watch the printers operate, with their buzzing and whirring, and then seeing the results of your work turn into a physical product. Additionally, for many people, the enjoyment of producing 3D CAD models themselves will be more popular than the actual physical printing process. You won't want to miss the next chapter, which walks you through the whole 3D printing workflow from idea, to CAD model, to print!

7/Overview of the 3D Printing Workflow

The previous chapters introduced you to 3D printing terms and various 3D printing options. This chapter will help you get your mind around how an idea becomes a 3D print. We will explore the processes and tools that we use in the rapid prototyping part of our business and explain how you can apply them at home.

You may have seen a 3D printer in action at a local technology event or retail store, or via an online video, and decided that you wanted to try it yourself. You signed up for an account and downloaded some free modeling software, as shown in Figure 7-1.

Figure 7-1. *A 3D modeler working with CAD software*

And...a couple of hours later you were frustrated with the uncooperative lump of digital clay on your computer screen. You still have positive feelings about 3D printing's potential but have experienced the disappointment of not being able to get the idea "from your mind to a design" on the computer. Don't worry. You wouldn't be the first person to have this experience.

3D printing is entirely within your reach, but you will need to approach it as you would any hobby or new technology and anticipate a learning curve. First, we'll look at the whole process (or workflow) in overview. Later, we'll look in greater detail at how to work successfully with that 3D "modeling clay."

3D Files

Before you can print anything, you have to have a 3D model that "tells" the printer what to print. The most common type of file format in the field of 3D printing is STL, which stands for "*ST*ereo*L*ithography." Don't be tricked, though; this file format is used by all 3D printers, regardless of type. The STL file simply defines points in space and connects those points together to form a series of triangles, which in total, are called a "mesh" or "mesh object." The file is called a "shell model" and can be thought of as a thin skin with a hollow interior.

These digital files are then sent to a software program, called a "slicer," which cuts a digital file into many, many tiny "slices" that instruct the 3D printer where to trace and lay down each layer to make the 3D object. Imagine taking a loaf of bread and slicing it into hundreds of horizontal slices. No two slices would be exactly the same—each would be slightly different from the ones right next to it. When they are all stacked one on top of the other, they end up re-creating a loaf of bread. If you have ever seen a medical MRI or CT scan, it's a similar process.

Future File Format

As of this writing, the industry standard is the STL file format. In April 2015, Microsoft and other prominent 3D printing companies formed a consortium to define an alternative file format that they claim improves on the STL format. It's called 3MF, and Microsoft said that it is designed to make 3D printing easier and more manageable. The new format would reduce loss of detail when exporting files for 3D printing, and it is designed to define vital information like color and material specifications. The 3MF file format is relatively new, but it has promising potential.

3D Models

You've seen 3D models all over the place! The latest science fiction and animated movies are filled with 3D models, like this one (Figure 7-2) from an open source film titled *Cosmos Laundromat*, which was created in free modeling software called "Blender."

Figure 7-2. *A still from Cosmos Laundromat ((CC) Blender Foundation (http://gooseberry.blender.org/))*

3D models are a digital representation of an object. Many of the 3D models created for movies and games were not modeled with the end goal of 3D printing. If you tried to download those models, the model probably won't be "optimized" for 3D printing and would need significant work in order to print. (We will discuss this in greater detail in Chapter 10, where we go over how to fix a 3D model in Meshmixer.)

Many of the real-world products we use in our daily lives begin their existence as 3D models, like the bike frame modeled in a computer-aided design (CAD) program called Fusion 360 shown in Figure 7-3.

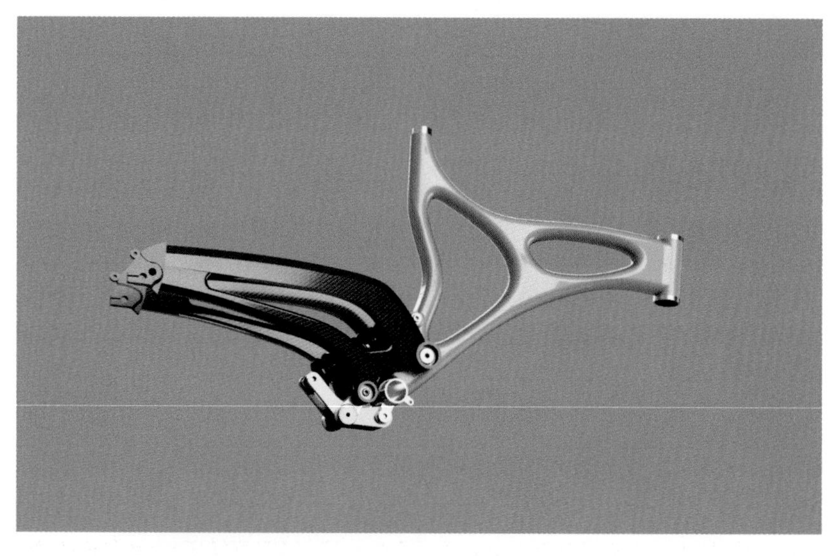

Figure 7-3. *A 3D model of a bicycle frame created in Autodesk's Fusion 360*

3D modeling programs are a great way to express yourself creatively and to bring your ideas into reality, and we will help you get started with them in the next section of this book. If you don't want to make your own, however, other people have created 3D models and have shared them freely online. You may want to start by printing out some of these prebuilt models. You'll find a more detailed description of where to find models and how to evaluate them in Chapter 8.

How Do I Get from Idea to Object?

In the rapid prototyping division of our company, HoneyPoint3D™, we tell customers that they can bring in an idea or sketch and we can turn that into a computer-aided design (CAD) file that will later be 3D printed. Many times, what they bring is a drawing on a piece of paper, sometimes even on the back of a napkin! Figure 7-4 shows the basic process of how we turn an idea into a 3D print.

Figure 7-4. *The workflow from idea to finished 3D print going clockwise from the left: 2D drawing, CAD file generation, 3D printed prototype, final piece*

Some clients come into our office with nothing more than an idea, but generally people have at least 2D drawings from which we can extrapolate measurement details. This rapid-prototyping process (described in Chapter 3) can be applied to creating anything from a princess figurine to a car engine.

Details of the 3D Printing Workflow

Here's a step-by-step guide for how to turn the idea in your mind into a physical object you can hold in your hand:

Idea
: Come up with an idea for an object you want to 3D print.

Search
: Thousands of 3D models already exist. Search on Thingiverse (*http://www.thingiverse.com*) or other online repositories for open source 3D printing communities to see if someone else has already designed it for you. If they have, download the STL file and proceed to the slicing step.

Plan
: If it doesn't already exist as a 3D file, draw your idea out on a piece of paper, with several different views. Draw your object from the top, the side, and the front. If you can, put in approximate measurements.

3D modeling
: Now that you have a plan, it's time to 3D model it in a software program. If you are a beginner to the CAD modeling world, we recommend you start with Tinkercad (*http://www.tinkercad.com*). If you are more experienced or want to create something more complex, then use Autodesk Fusion 360 (*http://fusion360.autodesk.com*). For organic shapes, use Meshmixer (*http://www.meshmixer.com*). Starter tutorials for all three CAD programs are provided in Chapters 9, 10, and 11.

Save and export
: In all of the previously mentioned programs there is an export function that will allow you to save your design in STL format. Save that STL file and download it to your computer.

Slice

Open up the slicing software recommended by your 3D printer manufacturer and load your model. It's best practice to use the program that your printer manufacturer recommends first, and then move to other slicing programs later on if you desire. Slicing an object creates many tiny horizontal slices that go through the entire object, as shown in Figure 7-5. These slices will later become the pattern by which the 3D printer knows where to trace and deposit material, for example, as with an FDM 3D printer.

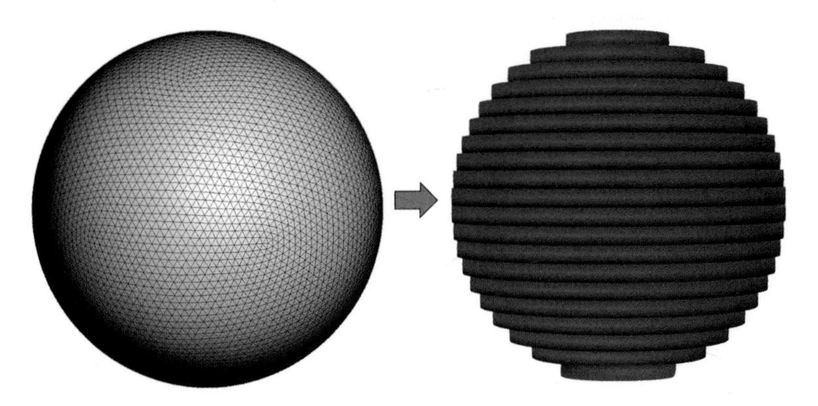

Figure 7-5. *Example of an object being sliced into many layers in order to get it "print ready"*

Evaluate support structure

Before you save your print file, you will use your slicer to determine if your model needs support structures to "hold up" overhanging parts. Support structures are needed to support parts of your model that would droop due to gravity during the build process if they were not supported. Support structures are meant to be a hand-removable scaffolding that the slicer will create for you automatically. See Figure 7-6 for an illustration of the original model and then the support structures added by the slicer.

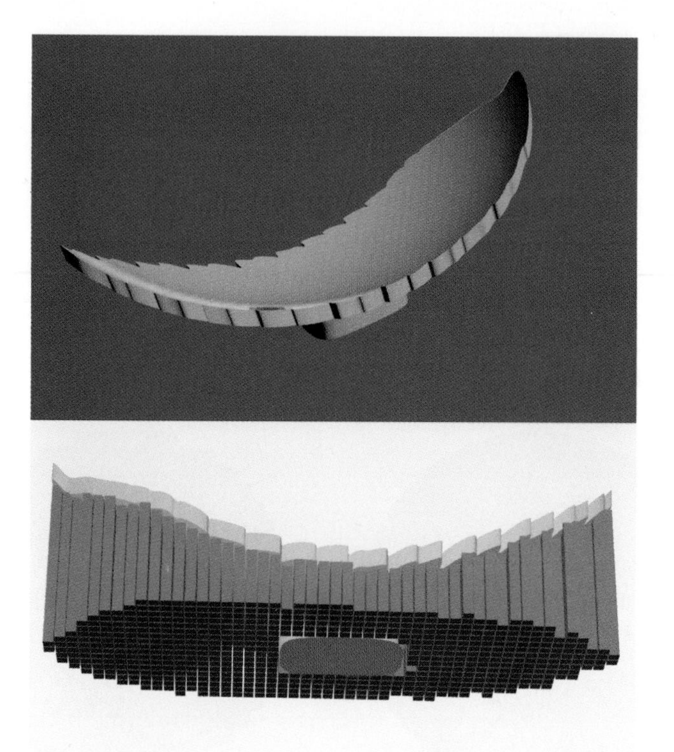

Figure 7-6. *(Top) A curved leaf on a pedestal. (Bottom) The curvature of the yellow leaf with automatic supports added in the slicer to make the overhangs printable. The supports (shown in orange) will later be removed by hand.*

Create G-Code/machine-readable file
: The slicer will create a digital file of printer-readable code that will direct the movements of the printer itself. Typically, this is a G-Code file. It may be named something else depending on the manufacturer of your printer, or it may be a series of *.png* images if you are using a DLP-based SLA printer.

Upload G-Code
: Upload the G-Code/machine file to the 3D printer from a direct connection on your computer, over a wired or wireless network, or to an SD card that will get inserted into the 3D printer.

Make sure your printer is ready

As we mentioned in Chapter 4, your print bed needs to be level to the nozzle. If your print bed is not properly calibrated, your print will fail. Most printers do not automatically calibrate themselves, so you will need to pay close attention to the manufacturer's instructions on how to level the printing bed. Manufacturer's instructions will also include how to properly load and store the materials you will be printing with and monitor the print while it's in operation.

Print

The big moment! Hit the print button and watch the magic happen!

Verify

Well, to be clear, it's not really magic that is happening. Make sure the first layer goes down evenly and well, and keep an eye on your printer throughout the build process. No 3D printer should be left unattended during the print process, and most 3D printer manufacturers will say this in their instructions.

Postprocess

After the print is complete, remove it by hand or with a small prying tool from the build plate, and remove any support structures by hand or with pliers. If you are using a resin printer, you will need to wash/clean your print and post-cure it for a few minutes in sunlight.

The workflow outlined here might seem daunting at first, but many of the steps do not take that long and will quickly become second nature to you. You'll feel a great deal of creativity and satisfaction once you have mastered the processes and begun making your own 3D prints!

You Can Outsource Any Part of the 3D Printing Workflow

Remember, though, you don't *have* to do it all yourself. You can outsource any part of the 3D printing process—even the idea stage! You can pick and choose which part of the workflow you want to do and leave the rest to someone else.

In the previous chapter we outlined the difference between 3D printing at home or outsourcing the job to a service bureau like Shapeways.com. If the process of 3D printing seems too much for you, outsourcing the printing part of the workflow is very easy and produces great results. (You might have to wait 10–14 days to receive your prints.) Figure 7-7 illustrates the difference between a 3D print made at home using a consumer FDM 3D printer versus using an outsourced service like Shapeways.com.

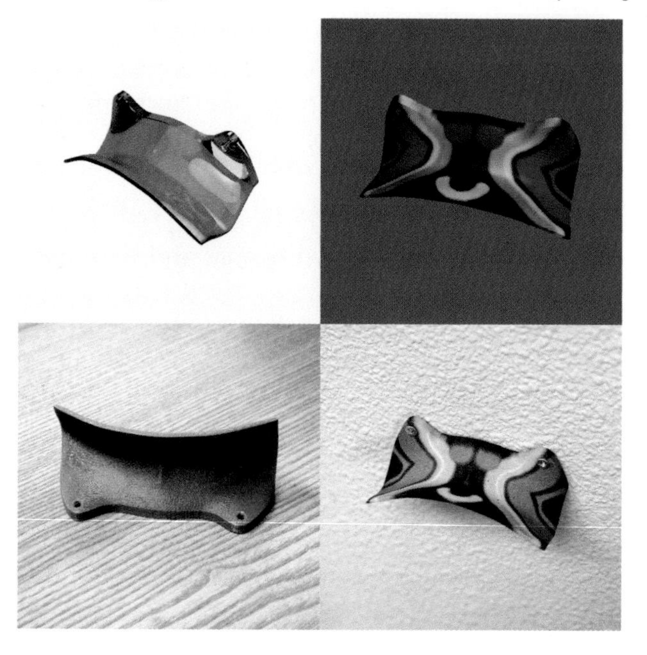

Figure 7-7. *A wall-mountable headphone holder. (Top left) A CAD model in Fusion 360. (Top right) Model colored in Meshmixer. (Bottom left) Model printed at home using PLA material. (Bottom right) The color 3D printed version made by Shapeways.com.*

So now you have learned about the steps needed to make a 3D print. The next chapter will go into more specifics of where to find models, what the 3D models are, and what to think about when creating your own.

8/Getting and Fixing 3D Models

In the previous chapter we looked at the procedures, or workflow, involved in bringing your 3D model from concept to reality. In this chapter we'll look at ways for you to get 3D models for printing and examine some of the issues you may need to consider when doing so. In the next chapter, we'll look at how to make your own 3D models from scratch.

Downloading a 3D Model

For most people who are getting started with 3D printing, it all begins with downloading an existing 3D model. In fact, if you purchase a 3D printer as a kit, very often there will be improvements to the 3D printer parts that the manufacturer will ask you to download and print. These downloads are often upgrades that the manufacturer has developed since the kit shipped. Such is the benefit of shipping a product that can effectively manufacture parts for itself! The model shown in Figure 8-1 is a fan shroud, which Printrbot asks you to print out for your printer that otherwise comes fully assembled.

Figure 8-1. *3D model of a fan shroud provided by Printrbot as a suggested first print*

Thingiverse (*http://www.thingiverse.com*) is a great place to find 3D models to download and print out yourself. Other sites are listed elsewhere in this book. Many of the models are free for you to download and print out for your own personal use. You can even modify them if you want to, using software tools we'll discuss in the next section.

3D Model Licensing and Legalities

Many of the 3D models you will find for download are provided under what is called a "Creative Commons" license (*http://www.creativecommons.org*). You should familiarize yourself with the licensing terms for any model that you download to print. The beautiful dragon shown in Figure 8-2 is an example of a user-submitted 3D model available for download from Thingiverse (*http://www.thingiverse.com/thing:600550*).

Figure 8-2. *Aria the Dragon (courtesy of Louise Driggers, used with permission)*

The Creative Commons license for Aria the Dragon states:

> This license lets others remix, tweak, and build upon your work non-commercially, and although their new works must also acknowledge you and be non-commercial, they don't have to license their derivative works on the same terms.

The Creative Commons license is very adaptable, and other models may come with different terms. On many places on the Web, including where you download models from online sites, the Creative Commons icons look like those shown in Figure 8-3.

License
Aria the Dragon by loubie is licensed under the Creative Commons - Attribution - Non-Commercial license.

Figure 8-3. *Creative Commons license with icons assigned to Aria the Dragon*

Technically, any 3D model can be converted into a file for 3D printing. But the word "technically" is used advisedly. 3D models created for movies or video games are often protected from copying by copyrights and trademarks. If you find a 3D model of your favorite character or item from Disney or Blizzard or some other entertainment company, make sure it's legal to print. Many entertainment companies today are, in fact, providing free models for fans to copy.

Creating 3D Models with Your Smartphone or Digital Camera

Not everyone is willing to put in the time it takes to become adept at the 3D modeling and rendering software that we'll introduce in the next section. Fortunately, there is an easy way to make 3D models with your smartphone or digital camera. The technique is called *photogrammetry* and is defined as the science of using images to create measurements. In this case, you simply take photos of the object you want to model and upload them to a service that will convert them into a 3D model for you.

The best-known and most widely used free photogrammetry service for consumers is called 123D Catch (*http://www.123dapp.com*). You can install it to your Mac or Windows PC or load it as an app to your smartphone.

To get started using 123D Catch, we recommend that you go to the website and watch the short "getting started" video. If it looks like something you want to try, sign up for a free Autodesk account at 123dapp.com. You'll want to find an object that you can walk all the way around and that is low enough so that you can take pictures of the top as well as the sides. Remember, you are creating a *three-dimensional* model—so you need images of the whole object. But don't worry about the bottom for now. Follow these steps:

- Take pictures of the object from all different angles and elevations, overlapping each picture by about 30 degrees.
- Make sure to raise and lower the camera as you take the pictures to get underneath any overhangs and to get the very top of your object.
- Most 123D Catch captures use around 30–60 photographs, so feel free to snap away!
- Copy the photographs to your computer.
- Open the 123D Catch application on your computer, load your photos, and give your project a name.
- Log in to your Autodesk account and initiate the model creation.

The 123D Catch application will email you when the 3D model has been created. Once the processing is finished you will be able to open the online viewer and see your 3D model. If you did not get all of the angles of your target object, the 123D Catch model will attempt to "fill in" missing areas. If large areas were missed, there will be a hole in your mesh.

Photogrammetry is an easy way of capturing organic shapes. While 3D modeling programs make it easy for you to create geometric shapes, the creation of realistic organic shapes often poses a challenge. Photogrammetry helps to solve this by translating complexity in the real world into low-to-medium complexity in the digital world.

If you want to 3D print something you created with photogrammetry, you will most likely have to fix the file before it is 3D printable. In this case, "fixing" means finding the holes in the models where information was not collected and "filling them in" to create a solid model. The work that needs to be done is more of the "cleanup" variety than anything that would require a deep knowledge of 3D modeling. We discuss this process in greater detail in Chapter 10.

Examples of What You Can Create In Photogrammetry

Models created in a photogrammetry application can be used as "starter geometry" and as a basis to build upon. Remember, in the 3D world, size does not matter, so you can create a model of something fairly small and make that 3D model look very large when you mix it with other objects. You can challenge people's assumptions about "what goes with what" in terms of size and function. Here are some ideas to help spur your creativity:

- Use 123D Catch to create a 3D model of a large seashell, and then 3D print that model as small earrings.
- Create a 3D model of an engaged couple and 3D print it as a wedding cake topper.
- Give an artist 10 pounds of clay and tell them to sculpt an object. Tell them to be as creative and detailed as they want. The resulting 123D Catch model can then be shrunk down and 3D printed in a smaller size to be, for example, a pendant.
- Capture a 3D model of an apple tree using 123D Catch, and then replace all of the apples with 3D models of sleeping cats (that you also captured with 123D Catch). This would probably be an Internet sensation!

One disadvantage of photogrammetry is the low level of detail the process captures, as shown in Figure 8-4. Even with a good DSLR camera the average detail level of a 3D model that comes out of 123D Catch is about 5mm to 10mm. Essentially, the features that are smaller than 5mm will just not show up on the resulting model. 5mm does not seem like a lot, but if you look at the transition area between someone's nose and their cheek you'll see it's less than 10mm. Photogrammetry's advantage is getting the general shape of an object easily and quickly.

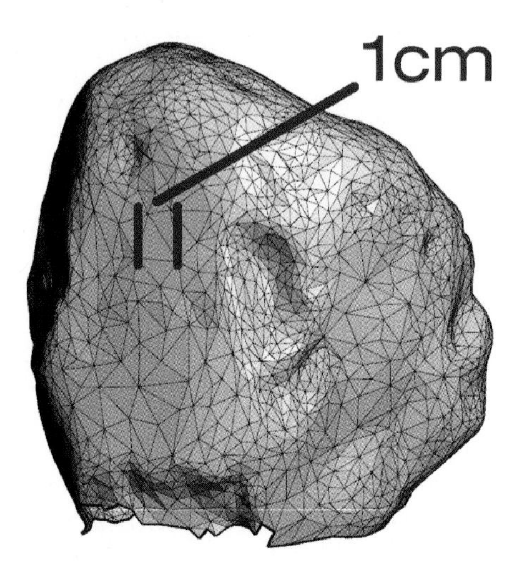

Figure 8-4. *A garden rock captured as a 3D model using 123D Catch, showing detail and, in some cases, lack of detail. A polygon can be as large as 1 cm.*

As with any technology, there are high-end and low-end devices. In our rapid prototyping division we use two different types of Artec 3D scanners to capture 3D images. They each cost around $20,000 and the phenomenal 3D scans they capture can be used for medical, art conservation, and other professional applications. Not everyone has access to a professional 3D scanner, though, and for most purposes a 3D model of average resolution obtained quickly and free of charge provides a great advantage.

Understanding 3D Model File Formats and Units of Measure

The file formats that define a 3D model are universal and can be opened by any number of applications. The major 3D printing file formats in order of popularity from first to last are:

- STL (Stereolithographic file)
- OBJ (Wavefront OBJ)
- VRML (Virtual Reality Modeling Language)
- AMF (Additive Manufacturing Format)

By a far margin, the most common format is the STL file, which describes nothing except the 3D shape itself. It does not provide texture, color, or volume information.

Printer Code

For the coders out there, here is an example of STL file syntax, which can be seen by opening the STL file with a text editor:

```
facet normal 0.000000e+000 -0.000000e+000 -1.000000e+000
  outer loop
   vertex  -6.065448e+000 -2.594533e+000 3.400000e+001
   vertex  -5.671127e+000 -2.966195e+000 3.400000e+001
   vertex  -6.199574e+000 -2.743562e+000 3.400000e+001
  endloop
 endfacet
 facet normal 0.000000e+000 -0.000000e+000 -1.000000e+000
  outer loop
   vertex  -6.199574e+000 -2.743562e+000 3.400000e+001
   vertex  -5.671127e+000 -2.966195e+000 3.400000e+001
   vertex  -6.292432e+000 -2.921260e+000 3.400000e+001
  endloop
```

The STL file really just describes three sides of a triangle in a 3D coordinate space, and then the next entry (triangle) starts off with one common side from the previous entry (triangle), and defines the next two sides of a new triangle as shown in Figure 8-5. This repeats over and over again, to create a "polygonal mesh" that all 3D printers know how to process. This file format is very simple and only contains information about the surface or "shell" of the 3D model.

The STL format also cannot contain any sort of color information for printing in color. (We added color to the mesh in Figure 8-5 to enhance the visibility of the triangles.) Nor does that file format have any information about the physical size of the object in the real world. A 3D model in an STL file will show as "x units high," but the physical unit of measurement is not described because size is mostly irrelevant to the modeling program. In an STL file, something that is 33 millimeters high looks the same as something that is 33 inches high. It is only when you want to print the model that the unit of measurement becomes important.

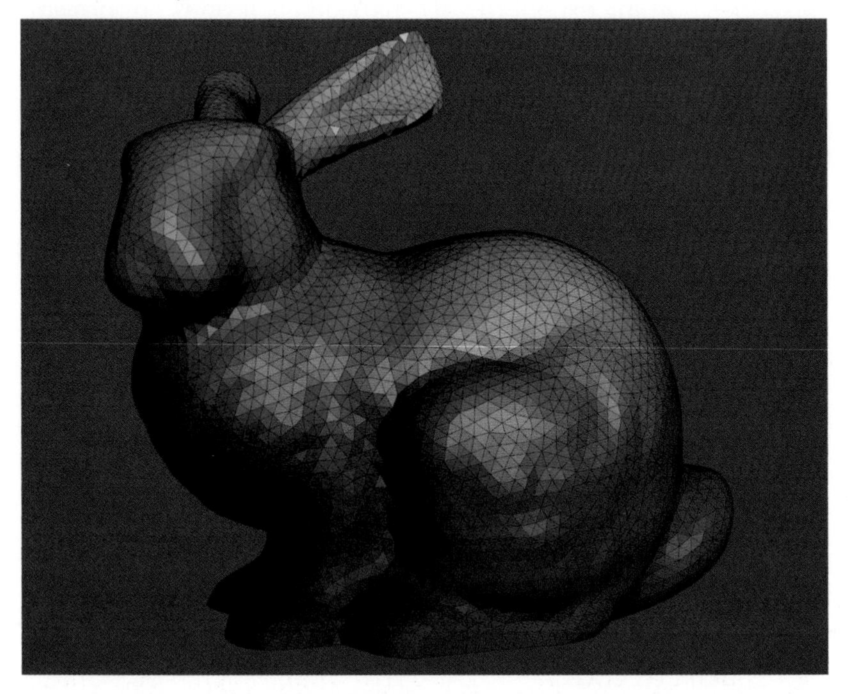

Figure 8-5. *Visual image of an STL file—a polygonal mesh, or shell, of a 3D model made of many triangles (10,996 triangles to be exact!)*

On a service bureau's website such as Shapeways.com, you will be asked to define the unit of measurement as shown in Figure 8-6.

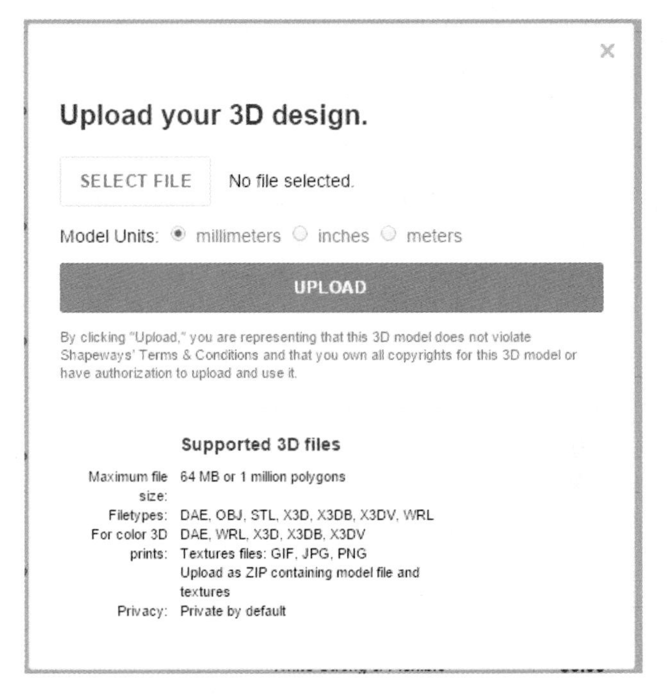

Figure 8-6. *File upload dialog from Shapeways.com prompting user to define unit of measurement. Note the maximum file size of 1 million polygons. The bunny had just over 10,000 polygons.*

Creating 3D Models with Four-Sided Polygons and Three-Sided Polygons

All 3D models you print with virtually all consumer 3D printers will need to be converted to the STL format, which is comprised of only three-sided polygons. We use the term "polygonal mesh" and not "triangular mesh" to describe most 3D models. A triangle is technically just a three-sided polygon.

There are other programs that can use four-, five-, six-, or more-sided polygons to create 3D models. A great example is the open source program Blender (*http://www.blender.org*), which has the ability to create multiple-sided polygons called "n-gons" in 3D models. There are benefits to using polygons with more than three sides; it gives the ability to create more graceful transitions between polygons, as opposed to sharp triangular points.

If your model's mesh is made from four- or more-sided polygons, you will need to convert that file into an STL (three-sided triangles) for 3D printing. This conversion from "more than three-sided polygons" down to "three-sided polygons" changes a model by adding in more geometry. This change usually works fine, but can sometimes create unexpected artifacts/changes on the new model that were not there on the starting model. For example, look at the perfect rectangle in the CAD model shown in Figure 8-7.

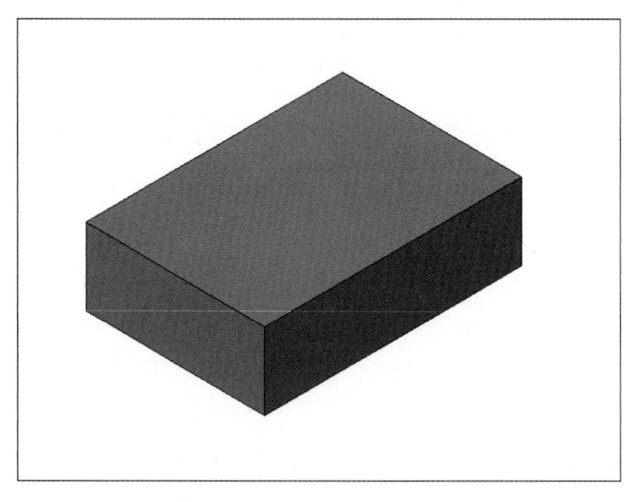

Figure 8-7. *A CAD-modeled rectangular object. Note the perfect rectangles on all sides (also known as "quads").*

When you export that model into a polygonal mesh editing program, in order to create an STL file from the CAD file, all of the four-sided polygons (quads) will be converted into three-sided polygons (triangles) as shown in Figure 8-8.

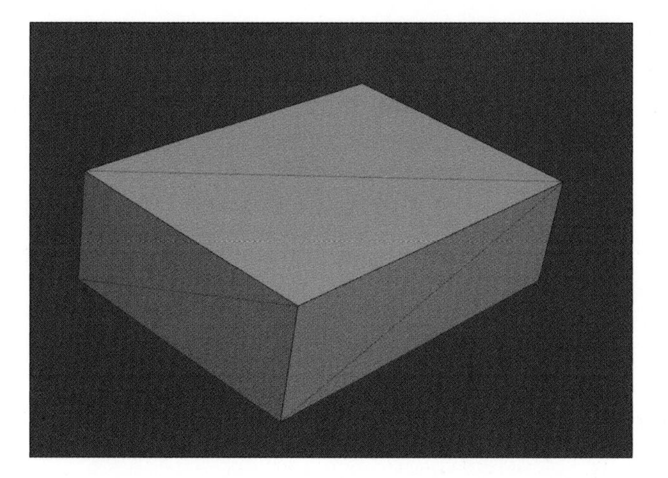

Figure 8-8. *A rectangular object exported to a three-sided polygonal format (STL). Note that there are no more four-sided areas. They have been converted to triangles.*

Fixing a 3D Model for 3D Printing

All models are not created equal. If you look through a repository like TurboSquid (*http://www.turbosquid.com*), which provides models for game developers, architects, and artists, you will find a lot of 3D models, and many of them are free to download. Since TurboSquid does not focus on models for 3D printing, however, many of the models you will find there will not print well. 3D models that are destined for use in computer animation or in computer games do not have to be as complete as 3D models that will be 3D printed. Even 3D models made with high-end 3D scanners and photogrammetry need some postprocessing to make them printable.

A common problem you will encounter when you have created a model using 123D Catch or a 3D modeling program is whether or not the model is "manifold." The term manifold refers to how "watertight" the model is. Think of it this way: all 3D models define a space that is either "outside" or "inside" the model, like the outside or inside of a ball. If you were to fill the inside of a manifold ball with water, no water would flow out. There would be no holes and no place for the water to leak through.

A nonmanifold model is the opposite. Water would leak out of a nonmanifold ball because there are holes, or—in the case of our model—missing data in the scan. Often nonmanifold models have large holes that are easy to see. Sometimes, though, the 3D model may *look* manifold but really is not. The sphere in Figure 8-9 looks manifold to visual inspection, but the software would find a problem.

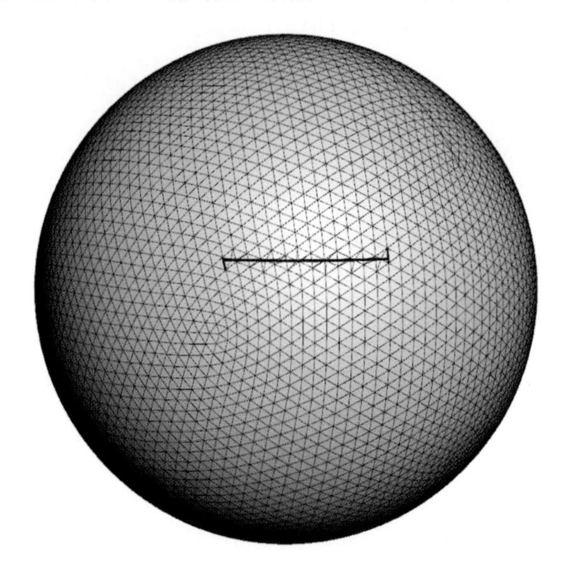

Figure 8-9. *A small cut in this model makes the model nonmanifold. If this were to be printed, the slicer would first have to make decisions about how to fix the model before printing.*

Using Meshmixer to Fix Your 3D Models

Without specifically using analysis tools, no amount of visual scrutiny, even if you zoomed all the way in, would show that there was a small cut in this sphere. The slicer software that would 3D print this model would have a problem. It's not a solid object it can print if there are holes.

Luckily there is hope! Most of the online 3D printing service bureaus offer tools to repair your 3D model for you. For the most part, these work well, but it is always better to have a good model going into the 3D printing process. That way, you are more likely to be happy with the end result.

Quite a few applications will help you repair models with the end goal of 3D printing. We believe a very good choice is Meshmixer, which we mentioned earlier and will describe in greater detail later. The Inspector tool in Meshmixer can often fix your model in just a few clicks, as shown in Figure 8-10.

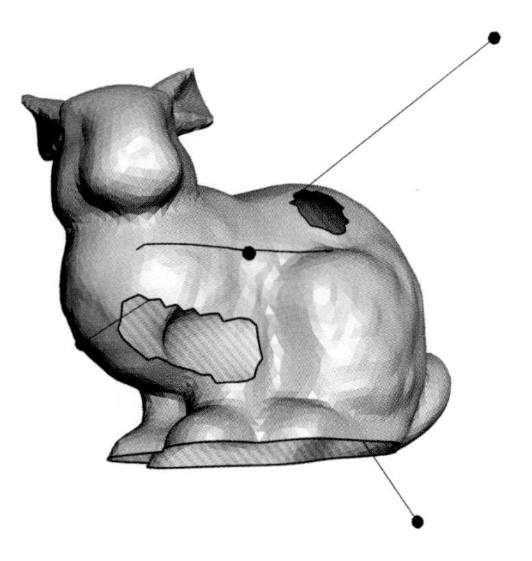

Figure 8-10. *The Inspector tool in Meshmixer showing problems with the 3D model*

Clicking any of the little bubbles coming out from the model will usually repair that part of the mesh successfully, making the model ready for printing. Because of this, Meshmixer is an ideal tool to integrate into your existing workflow for 3D printing, even if you are already comfortable using other tools.

Meshmixer occupies a unique position in the 3D printing world. It does an excellent job of bridging the gap between the digital world of 3D modeling and the physical world of 3D printed objects. No other software program has such comprehensive tools to help you get a physical object created. We'll go into greater detail on how to get up and running with Meshmixer in Chapter 10, and we will describe how to use Meshmixer to fix problems with your 3D model.

PART III

CAD Tutorials

9/Getting Started with Tinkercad

In the next three chapters we will take you step-by-step through the basics of creating a 3D model from scratch using three CAD modeling software programs: Tinkercad, Meshmixer, and Fusion 360. You can view them respectively as beginning, intermediate, and advanced programs, though we have taught students as young as nine years old to use Fusion 360.

These CAD programs have the benefit of being free (or near free), so trying them out is a great place to start. These beginning tutorials are short enough that you should be able to complete each one in an hour, but you should feel free to move through them at your own pace.

Tinkercad (*http://www.tinkercad.com*) is an online CAD modeling platform from Autodesk that is free for anyone to use. It's known as a quick and easy way to create 3D models and is simple enough for beginners and kids to pick up easily. In fact, in our educational courses, we always start beginning students, regardless of age, on Tinkercad. See Figure 9-1 for an example of models created in Tinkercad.

The platform is entirely browser based, which allows for your work to be instantly saved to the cloud and available on any other computer. Tinkercad runs best in modern browsers such as Firefox, Safari, or Chrome. Internet Explorer has been known to work, but the experience is more fluid with the other browsers mentioned. Keep in mind that slow Internet speeds can negatively affect the overall experience.

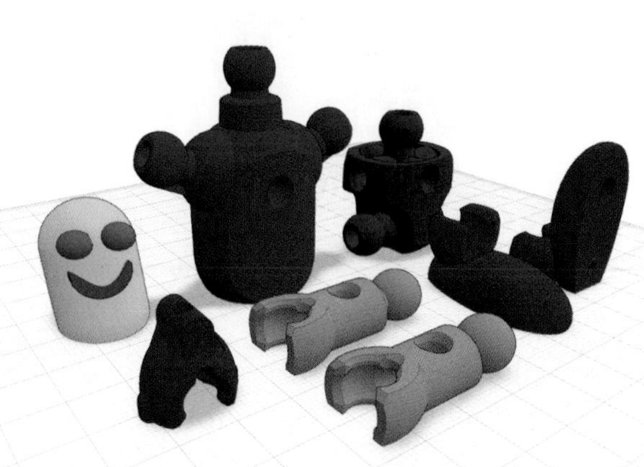

Figure 9-1. *Action figure created in Tinkercad, using "Tinker-play" components, all 3D printable*

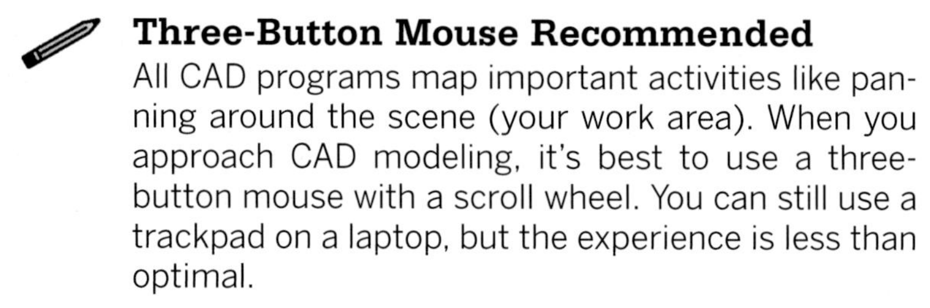

Three-Button Mouse Recommended
All CAD programs map important activities like panning around the scene (your work area). When you approach CAD modeling, it's best to use a three-button mouse with a scroll wheel. You can still use a trackpad on a laptop, but the experience is less than optimal.

Setting Up an Account

In order to access Tinkercad you will need to have an Autodesk account. The good news is that this program is free and the promotional emails you might receive from Autodesk are minimal.

If a child under 13 years of age is signing up to use Tinkercad, their user ID request will get routed into a COPPA (Children's Online Privacy Protection Act) compliant subsystem inside of Autodesk that requires an adult to verify that the child is allowed to create an account (Figure 9-2).

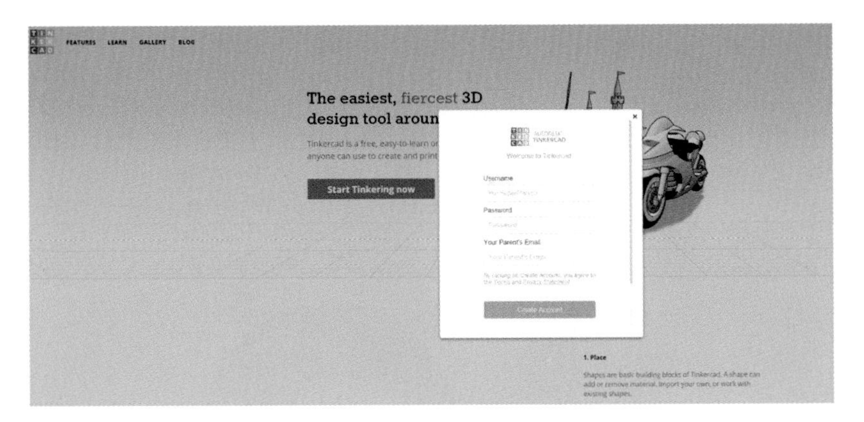

Figure 9-2. *Tinkercad account signup requiring parental permission for users under 13 years of age*

The child's account will be in a special part of the Autodesk infrastructure where it will comply with the COPPA requirements. There will be restrictions on the sending of marketing emails as well as restrictions on sharing that child's account information with third parties. This is a good thing for the protection of children, but keep in mind this extra step can add up to 24 hours of waiting time for kids to join the Tinkercad community. This is especially important to remember if you are using Tinkercad in an educational setting where parents will need to pre-register their kids on the platform.

Before You Pick Up the Mouse

We recommend you first use pencil and paper to draw your idea out. The basic premise for all CAD design is understanding that shapes interact with other shapes to create new shapes. This concept sounds easy enough, but if you are just starting to learn CAD, the best place to start is having a solid understanding of what you are trying to create. For most people, drawing it out on paper is more intuitive. Don't worry if it's not to scale or if your drawing of a house looks more like a shoebox. You will benefit greatly by having this visual guide as a reference.

If you draw a circle on a piece of paper, and then you draw another overlapping circle just touching the side of that one, it looks like a figure "8." This is an example of "additive" construction—adding two things together to create something new. Conversely, if you draw a circle and then draw a smaller circle inside it, and "remove" the inner circle, you end up with the letter "O." This is subtractive construction, i.e., building something new by taking *away* material. Both examples are shown in Figure 9-3.

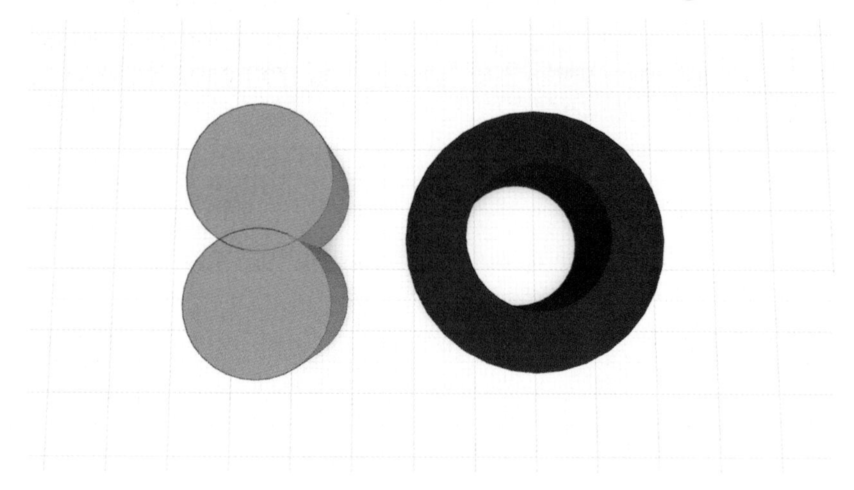

Figure 9-3. *Line drawings showing additive and subtractive properties*

"Boolean" Defined

The important CAD term "Boolean" refers to the addition or subtraction of shapes from other shapes to accomplish an end goal. Tinkercad is adept at performing Boolean unions (adding) and Boolean differences (subtraction), which are the core of how to start modeling your design.

In the following tutorial, we'll use Tinkercad to model a simple nameplate. This tutorial will go over three key procedures: adding shapes on the "Workplane," turning shapes into "holes" or "cutting tools," and downloading your model for 3D printing.

Creating Shapes in Tinkercad

When you first start Tinkercad we highly recommend you go through the software's free set of online tutorials. These great lessons will walk you through Tinkercad's features at a pace you can set for yourself. They are available at the top of your Tinkercad account page under the "Learn" link. You can always click the Tinkercad logo shown in Figure 9-4 located at the top left to return back to your main account page.

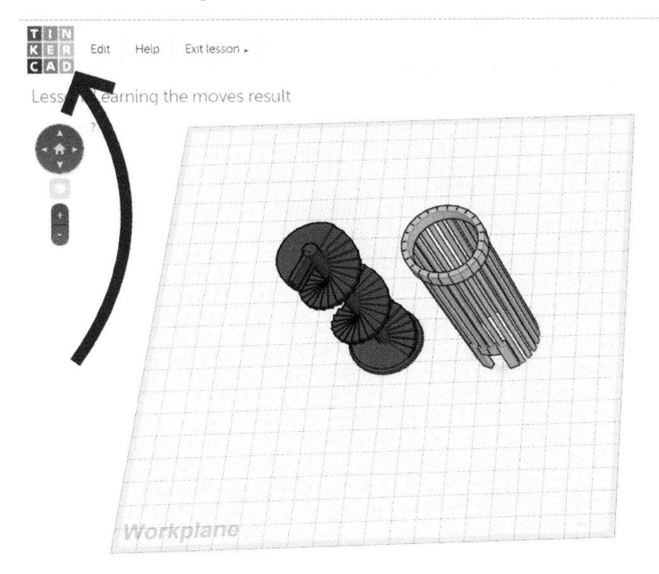

Figure 9-4. *The Tinkercad logo takes you back to your home page*

When you return to your home page, click the "Create new design" button (Figure 9-5) to start a new design.

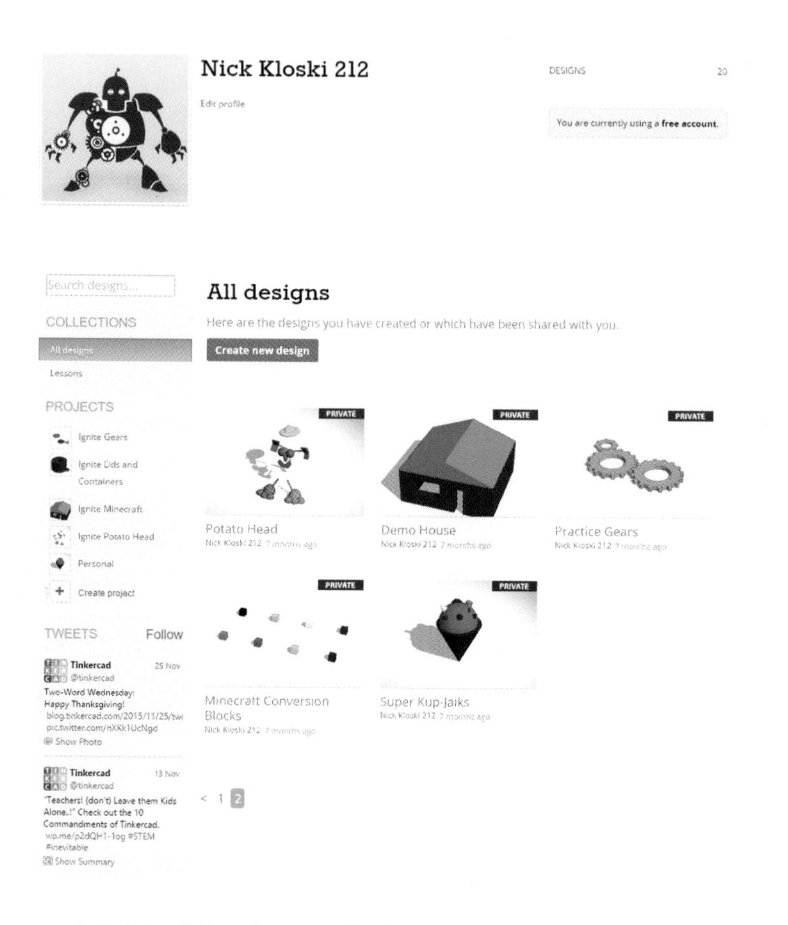

Figure 9-5. *The "Create new design" button lets you begin your new project*

You are presented with a blank canvas (Figure 9-6), upon which you will create a masterpiece! Or, as is true with most model design, you will create something that you will call "your first try" and then figure out how to make it better over time.

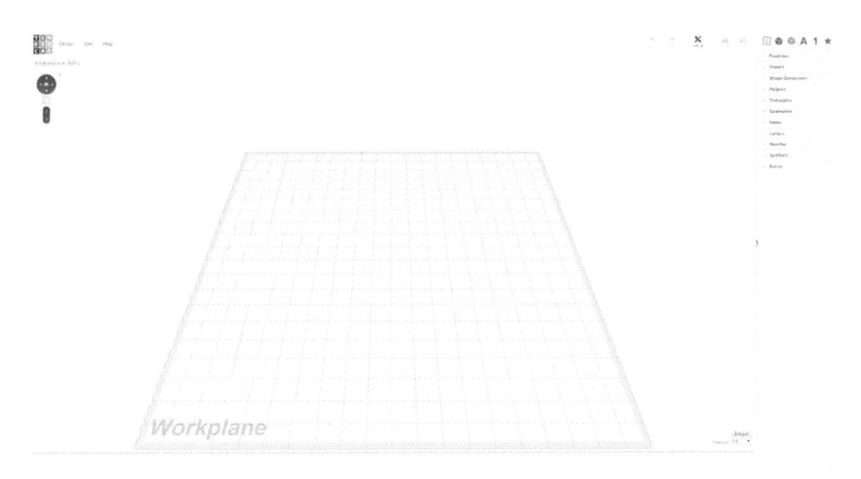

Figure 9-6. *The blank canvas that is called the Workplane*

Next to the Workplane you just saw, on the righthand side of the screen there is a menu bar filled with premade shapes you can use. These shapes will help you create your 3D model for eventual 3D printing.

Click the Geometric section in order to drop down the menu of shapes, as shown in Figure 9-7. Click and hold your mouse button down on the red "box" and drag it out onto the Workplane.

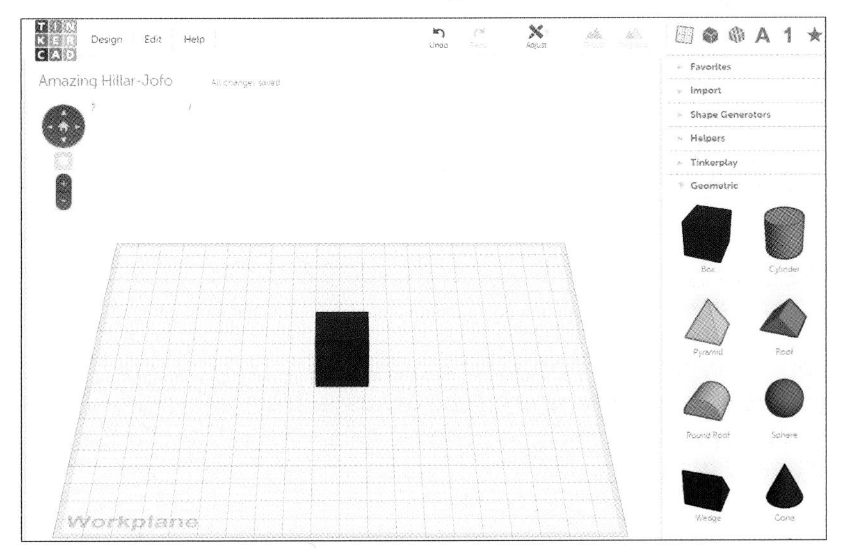

Figure 9-7. *You can click and drag shapes onto your Workplane*

You now have a basic shape you can adjust. If you click your red box and hover your mouse over the small white squares at the corners, dimensions will pop up as shown in Figure 9-8.

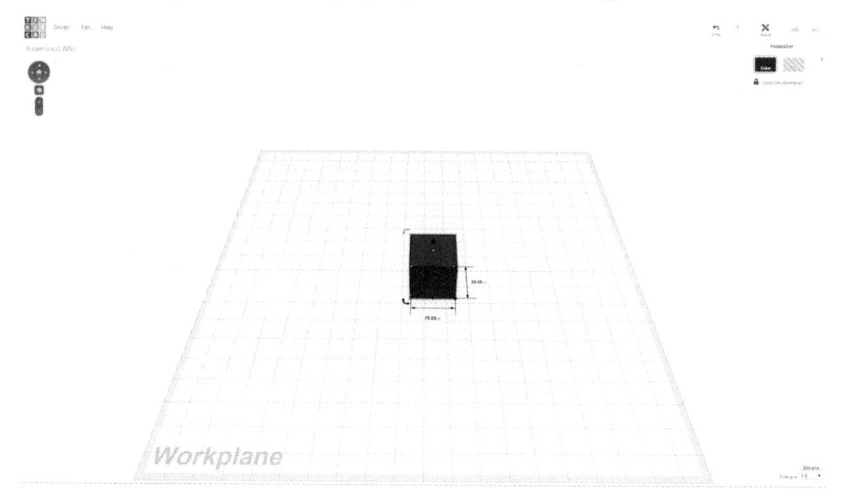

Figure 9-8. *After clicking the box, small white squares appear at the corners, allowing for resizing*

The default dimensions of the box are 20mm wide by 20mm deep by 20mm tall. Clicking and dragging those small squares will resize the cube in that particular direction. Carefully roll your mouse over a tiny white square, and it will turn red to indicate that you can click-hold it. Sometimes it is hard to really click that small white square, so have patience!

These tiny white boxes and other shape modifiers are called "handles," and they allow you to resize and reshape any object in Tinkercad. See Figure 9-9.

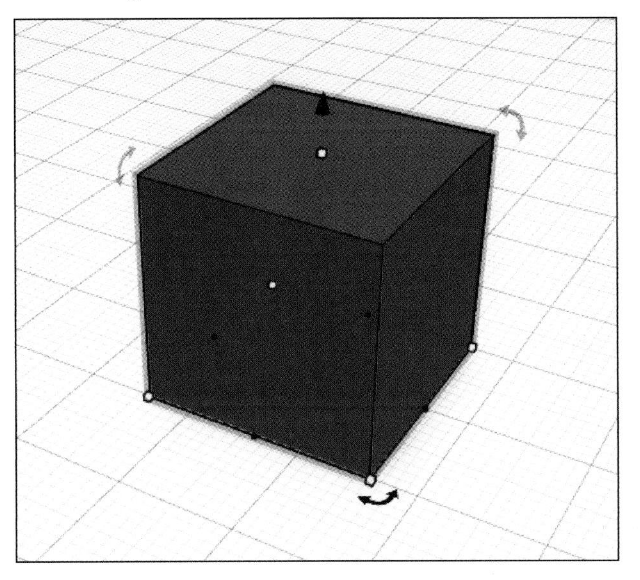

Figure 9-9. *Various handles on an object in Tinkercad*

Here are descriptions of the various handles on your cube:

- Four white boxes on the bottom corners to resize the shape in those directions
- Four black dots on the midlines between the white handles that allow you to resize the shape by the edges instead of the corners
- A white box on top of the object to make the shape taller
- A black arrow on top of the object to physically raise the object off of the Workplane it is sitting on
- Three rotation arrows that allow you to rotate the shape in any of the three directions (also note that you might not be able to see one of the rotation handles until you rotate your view around to the side of the box)

Rotating an Object

There are two ways to rotate your object: by using your mouse or by using the application itself. If you look in the top lefthand corner of the main Tinkercad screen, you will see a disc with movement controls as shown in Figure 9-10.

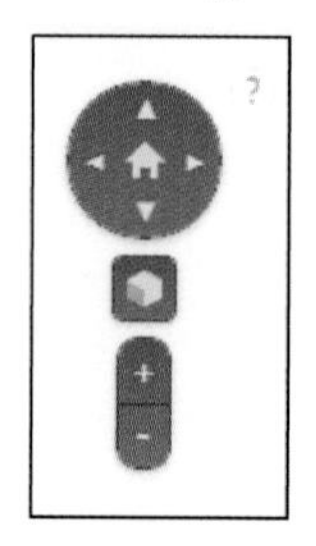

Figure 9-10. *The navigation panel in Tinkercad*

Clicking the arrows inside the "home" disc allows you to rotate your object. Clicking the plus and minus (+ and –) signs allows you to zoom in and out. Select an object by clicking it. The 3D "box" symbol below the disc zooms your view into that object so that it takes up the entire screen.

Helpful Shortcuts

If you want a keyboard shortcut to zoom into an object, just click the object and hit the "f" key to "focus" and the object will fill your screen. If you want to get back to the default view, you can click the "home" icon (in the disc) and your view will be reset. In general you can navigate more easily with your mouse than your keyboard.

If you click the small question mark icon next to the home disc, you will get a helpful pop-out window as shown in Figure 9-11.

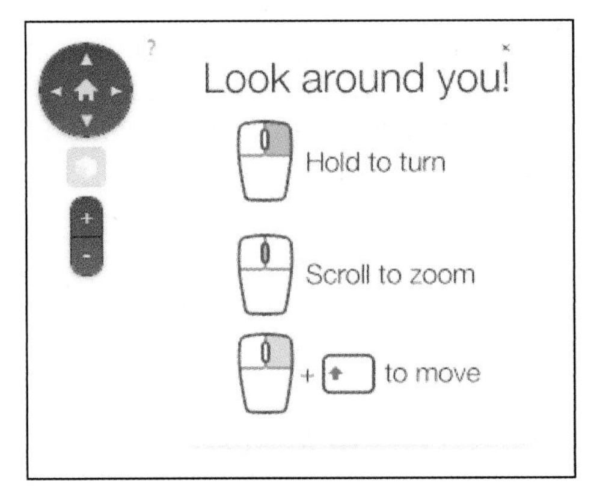

Figure 9-11. *Expanded navigation help menu on Tinkercad*

Holding down the right mouse button will allow you to pan the scene to see your design from different directions. Zooming is accomplished by the mouse wheel. If you want a centered view on a different part of your Workplane, then hold down the Shift key while right-clicking to pan the scene.

Changing the Shape

Now that you have the basic movement down, it is time to start creating your idea! If you resized the original red box by clicking and dragging on the handles, you will have noticed that the measurements have changed. These measurements can be directly sent to a 3D printer and will be 3D printed accurately!

Resizing the Red Box

Resize the box to dimensions of 5mm high by 40mm wide by 60mm long, as shown in Figure 9-12.

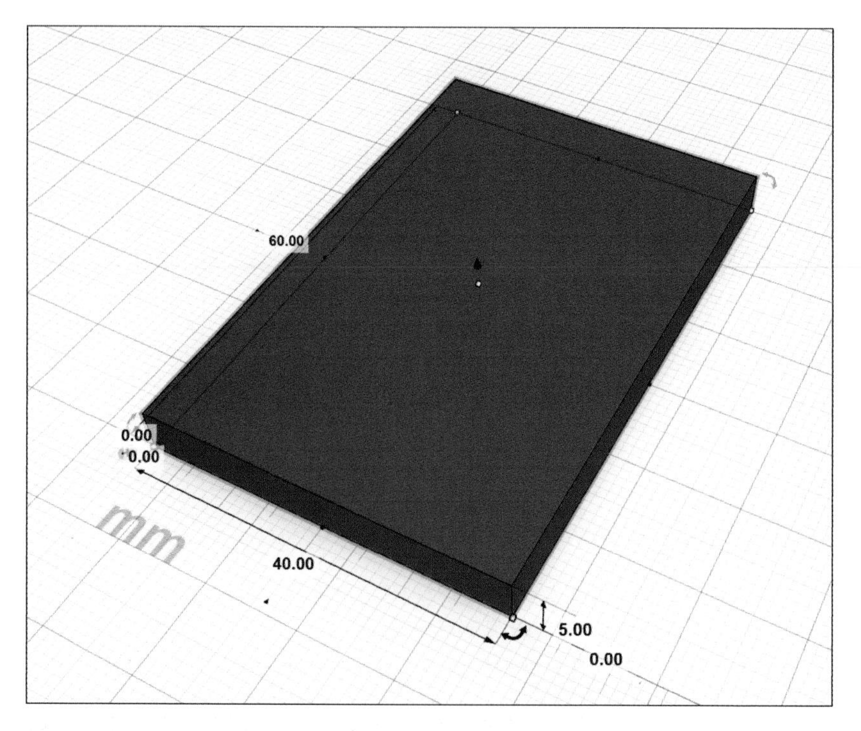

Figure 9-12. *Resized original red box using handles*

You now have a flat rectangular base. We'll use this as an example of how to use shapes "negatively" as cutting tools to create new shapes. This is a key concept not just in Tinkercad but in all CAD programs.

In the righthand menu bar list of tools, scroll down to the "Symbols" toolbox and drag a heart shape onto the Workplane. Anywhere is okay for now.

Click the heart shape and make sure it's selected before you perform these steps:

- Using the tiny white box handles, make it 10mm high.
- Use the rotation arrow in the navigation tool to rotate it 90 degrees counterclockwise. Your Workplane should look like Figure 9-13.

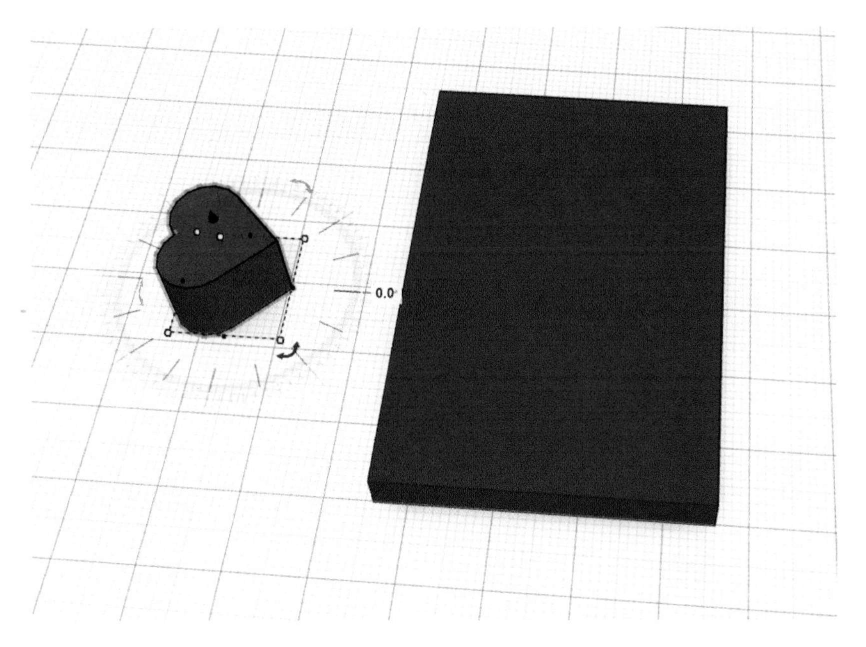

Figure 9-13. *Stock heart shape rotated to a specific angle*

As you click and drag on the rotate handles a grid will appear to help you find the right degree-angle. As you are holding down the left mouse button and rotating the object (if your mouse is outside of the rotation circle) you will move in single-degree increments. If your mouse is on the inside of the rotation circle, your object will jump in 22.5-degree increments. Or in other words, by these increments: 22.5, 45, 67.5, 90, etc.

You can rotate your view so that the heart is right-side up. Drag your heart inside the rectangle near the corner, as shown in Figure 9-14.

Figure 9-14. *Positioning the heart in the upper-left corner*

Now comes the fun part! Click the heart to select it. In the top right of your screen look at the window entitled "Inspector," as shown in Figure 9-15.

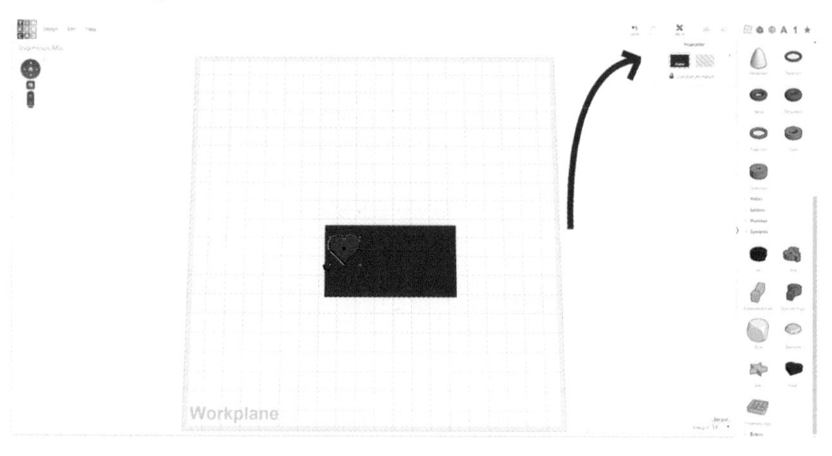

Figure 9-15. *Inspector window when an object is selected*

Clicking the left box will allow you to change the color of the pieces you selected, but for this step, click the "hole" box (shaded gray strips) and see what happens to the heart!

A Note on Color

Though you can change the color in Tinkercad, this is only a visual aid for you. The end file that you download for 3D printing will be in STL file format, which does not have the ability to convey any sort of color information. The final color of your object will be whatever color filament or 3D printer feedstock you are using.

The heart has now become gray striped to denote that it will be used as a cutting tool. Your heart will look like the one shown in Figure 9-16.

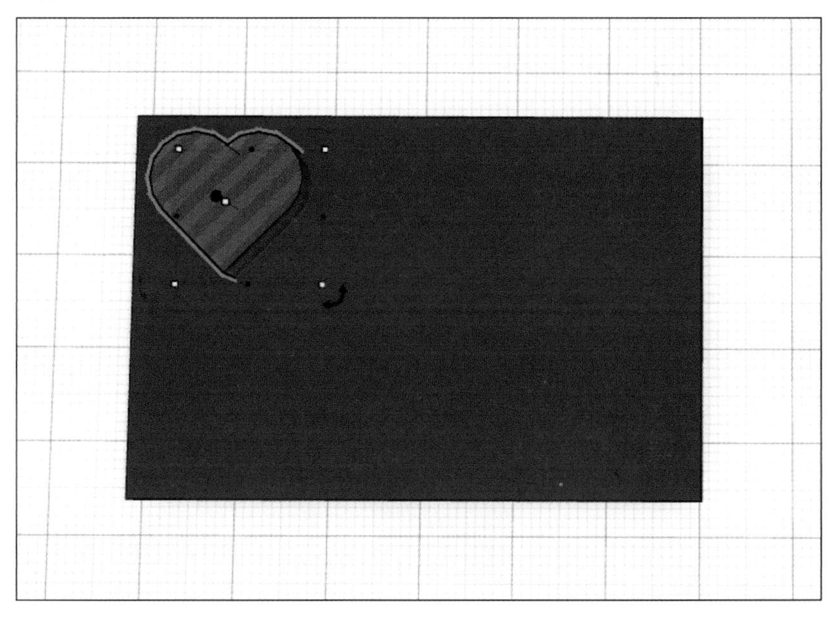

Figure 9-16. *Heart changed to "hole" to denote it is now a cutting tool. You are one step away from creating your first subtractive shape!*

Grouping and Ungrouping Shapes

Tinkercad is a "parametric" tool, which means that behind the scenes, everything you are interacting with on the screen is actually based on simple mathematics. The "parameters" that you set for length, width, and height determine the shape of an object. Also, Tinkercad has the concept of "nesting" or grouping objects together. Objects that are grouped together have special abilities that nongrouped objects do not.

Grouping objects in Tinkercad saves you a lot of time in that you can move all of the grouped objects together all at once. And here is another great feature of the grouping: you can create a cutting tool (like the heart), then once you group the objects together, you will actually see the effect of the cutting tool in your model on all grouped parts!

Let's try the grouping function. In the upper-left corner of your Workplane, click and hold your left mouse button and drag a box around both the heart and the rectangle, selecting both of them.

In the top right of your screen, click the "Group" button as shown in Figure 9-17.

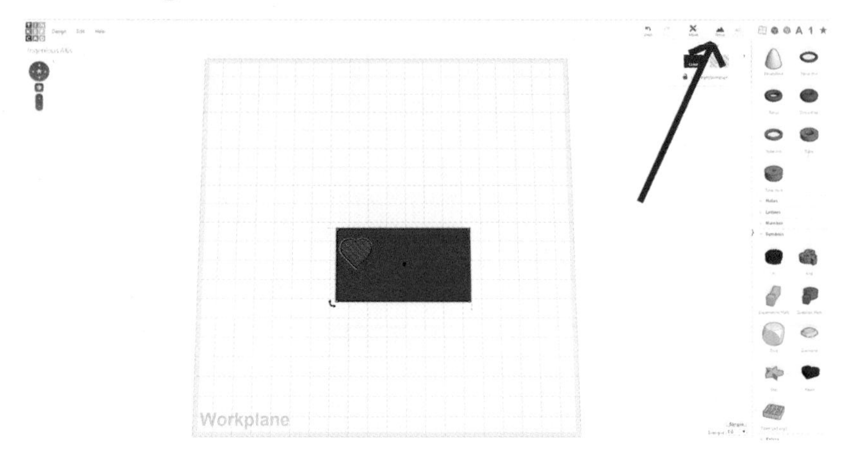

Figure 9-17. *Location of the Group button*

You will see a result that looks like Figure 9-18.

Figure 9-18. *Result of grouping heart cutting tool and rectangle*

The heart is now cut out of the rectangle! With Tinkercad being a parametric modeling program, this also means that the operation you just completed of cutting the heart out of the block is "non-destructive." That means that you can go back in to change things around. This will not disrupt other objects in your design. It's like it has a memory for what you did in layers.

Double-click with your left mouse button on the rectangle and "enter" into the group you just created. You can change the placement of the heart, or really any component that is in that group. When you click outside of the grouped shapes, the various cuts are performed based on the new locations of the shapes. Conversely, you can also click the "Ungroup" button to ungroup selected shapes.

Finishing Your New Creation in Tinkercad

The next steps of this tutorial will demonstrate how to further customize your heart-plate using other customization tools:

1. Go to the Shape Generators to the right of your screen and expand the menu bar to find the "Text" object tool as shown in Figure 9-19.

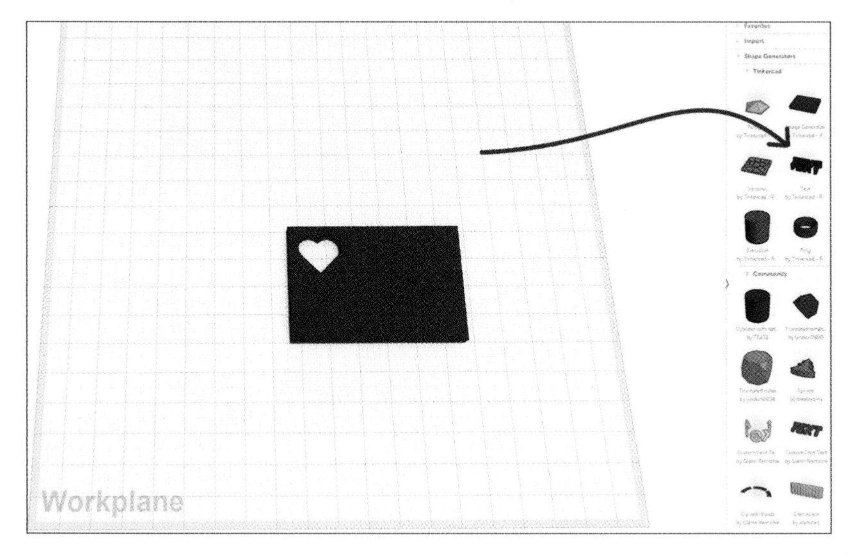

Figure 9-19. *Find the Text object tool*

2. Drag the Text object out onto the Workplane as shown in Figure 9-20.

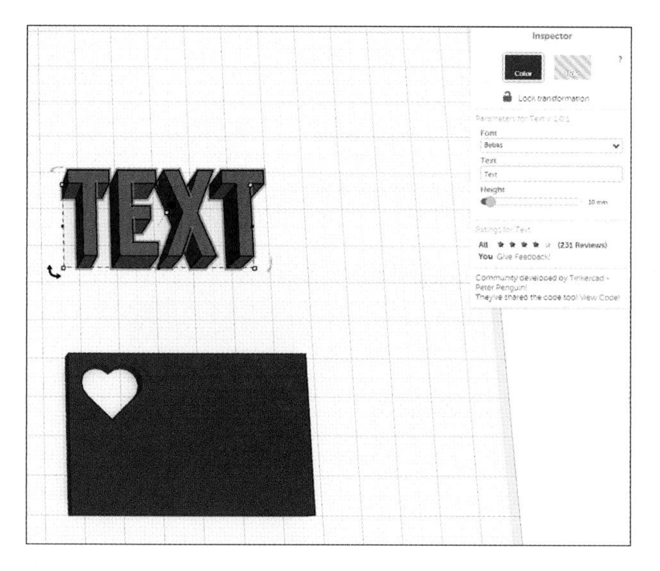

Figure 9-20. *Text object on the Workplane*

3. Use the Inspector tool in the top right to change the "Text" to something else as shown in Figure 9-21.

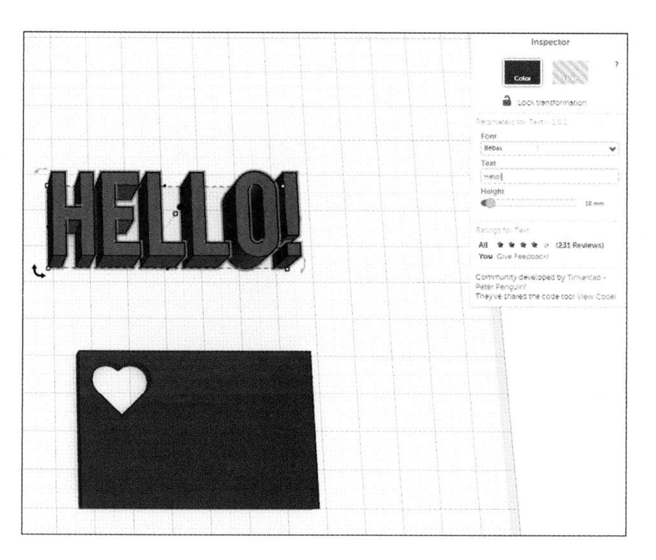

Figure 9-21. *Text changed to "Hello!"*

4. Drag the text into your nameplate (the rectangle) and resize if necessary as shown in Figure 9-22.

Figure 9-22. *Resized "Hello!" placed on the nameplate*

5. With the text "Hello!" still selected, use the Inspector window in the top right to turn the Text object into a "hole" as shown in Figure 9-23.

Figure 9-23. *"Hello!" turned into a "hole" cutting tool*

6. Using your mouse, drag a box around all of the objects. Next, in the top right of the Tinkercad window, click the Group but-

ton at the top right of the screen to see the result shown in Figure 9-24.

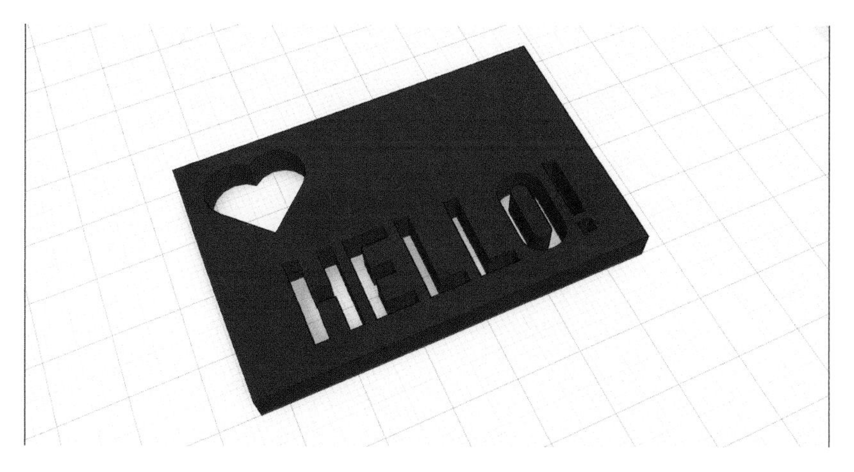

Figure 9-24. *End result of cutting after objects are grouped*

7. Click "Design" in the top left of the Tinkercad window as shown in Figure 9-25.

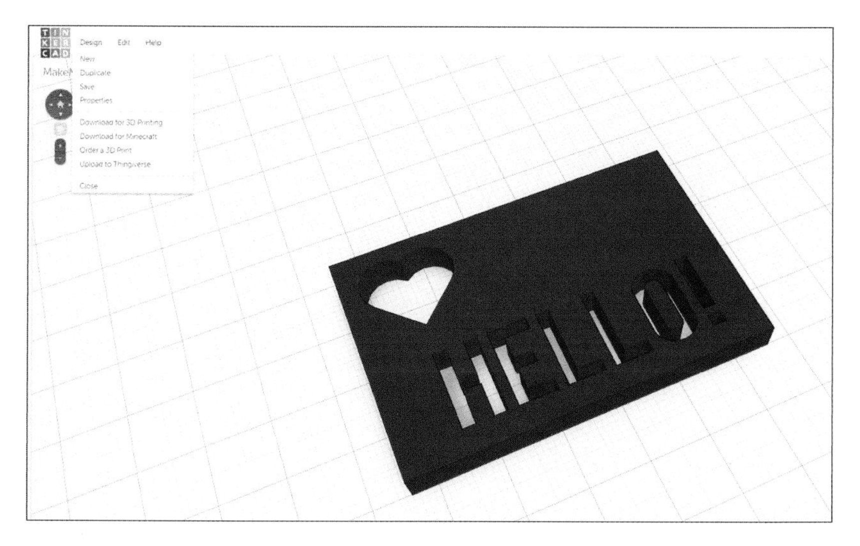

Figure 9-25. *Drop-down menu when clicking on "Design"*

8. Click "Download for 3D Printing," and then choose ".STL" as shown in Figure 9-26. This is the file format you could 3D print!

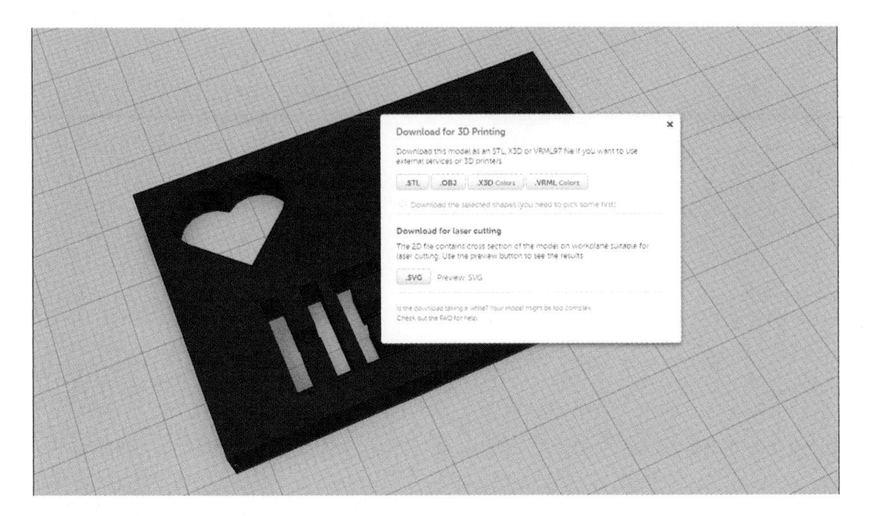

Figure 9-26. *Saving file as STL from the Design menu*

Congratulations! When you have an STL file from Tinkercad, you can load that file into your 3D printer or upload it to an online service bureau where they will 3D print your design for you.

Now that you understand the basics, we recommend that you spend time playing around with Tinkercad on your own until you feel comfortable with the features and controls. You will learn best by doing, so don't be afraid to experiment.

What's Next?

In Chapter 10 we give you another CAD tutorial. We'll look at Meshmixer, another fun and free CAD modeling software program from Autodesk. This one has a medium level of complexity but allows you to do far more with the software than what you can do with Tinkercad.

If you already have Meshmixer installed on your computer, you can double-click (or import) the STL file you just created into Meshmixer for further editing, as shown in Figure 9-27.

Figure 9-27. *The STL file loaded into Meshmixer*

Some changes are needed to make the file more easily editable in Meshmixer. You can do this by "remeshing" (using the Meshmixer commands Edit → Remesh → Linear Subdivision a couple of times) to make the model more complex so that Meshmixer has enough triangles to work with. But, once the model has enough triangles in it, you can do some fun things in Meshmixer (Figure 9-28).

Figure 9-28. *Free-form sculpting and modeling in Meshmixer*

When you're ready for a Meshmixer tutorial, just turn the page!

10/Getting Started with Meshmixer

Meshmixer is another free program, and it is amazingly powerful. It's so useful that, if you 3D print a lot, you may find yourself using Meshmixer before every print. Don't be fooled by the free price and by our medium-level skill rating; it can be just the right toolbox beginners need to print and create models as well as something advanced users can use to create digital masterpieces. Check out the model in Figure 10-1 for an example of the kind of sculpted 3D model you can create with Meshmixer.

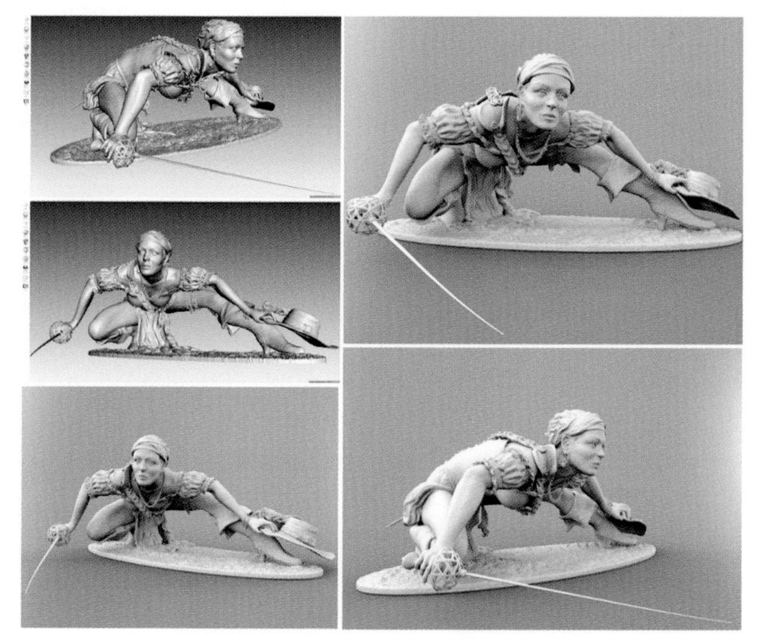

Figure 10-1. *The two gray figures (top left) are from Meshmixer; the other three models have had shading effects added in an external program ("Madlon" by Gunter Weber/"MagWeb" (http://www.meshmixer.com/forum))*

Meshmixer has tools for the creation of 3D models, as well as tools that will help you successfully 3D print them. In this way, Meshmixer is a unique product because it sits between the digital and physical worlds. Here are a few reasons why Meshmixer is one of our most popular online 3D printing and 3D modeling courses on Udemy.com:

- Meshmixer can be used not only to create all kinds of 3D models, but also to help you translate those models into physical products. This ability is unmatched by any other software program.
- Even if you never end up 3D printing what you create, Meshmixer is a powerful standalone digital creation tool.
- Meshmixer helps you get the most out of your own 3D printer at home.
- Students, professionals, artists, hobbyists and designers, and beginners and advanced modelers can all benefit from learning Meshmixer.
- Meshmixer can be taught to ages 9 and up.
- There are over 80 tools in the program, many of which you can start using right away.

Getting Started

Meshmixer is not browser-based like Tinkercad, so you will need to download the program by visiting the official website. Go to http://www.meshmixer.com and follow the instructions to download and install Meshmixer on your computer.

When you first start Meshmixer you will be presented with a menu of starting choices, as shown in Figure 10-2.

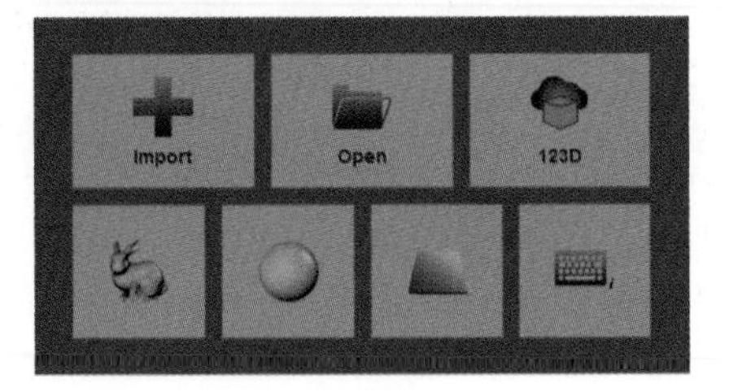

Figure 10-2. *Meshmixer Start menu*

Before we get any further, let's go over some basic aspects of Meshmixer:

- Left-clicking your mouse will select things on the screen, or act as the brush when sculpting
- Holding down the right mouse button and dragging your mouse will rotate the object
- Holding down the middle mouse button will pan the object

Figure 10-3 shows the lefthand side of the Meshmixer screen with labels of the tool functions. You are able to access all the tools of Meshmixer from these main icons.

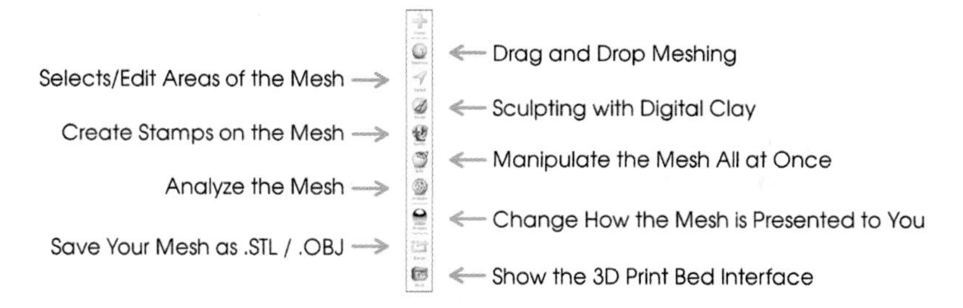

Figure 10-3. *Meshmixer Tool menu and descriptions*

Starting the Tutorial

From the Start menu, click the rabbit in the lower-left corner. This will load up the iconic "Stanford Bunny," which was 3D scanned from a small statue in 1993 at Stanford University. Once you click the icon, a bunny will appear on the screen. You now have a 3D model to work with.

Now that you have the bunny on the screen, though, you will need to "fix" it. Because this is a 3D model that was created by 3D scanning a real-world object, the scanner was only able to capture the exposed surface areas of the bunny, and not the parts where it was sitting on the table. To 3D print this bunny, this "missing and open" part of the scan will need to be closed. 3D printing requires a closed, airtight model. Fortunately, Meshmixer has some powerful tools that will evaluate and "fix" 3D models to make them more 3D printable.

Begin by clicking Analysis → Inspector.

You will see a blue "pin" popping out of the bottom of the bunny as shown in Figure 10-4. This indicates that there is an issue with the mesh in that location. You can easily see the big hole in this 3D model, but some holes are harder to see, making this tool very valuable. Meshmixer will find errors such as holes and small disconnected parts with the Inspector tool. Make sure to rotate your model all the way around with your right mouse button to see all pins that might be hidden from view.

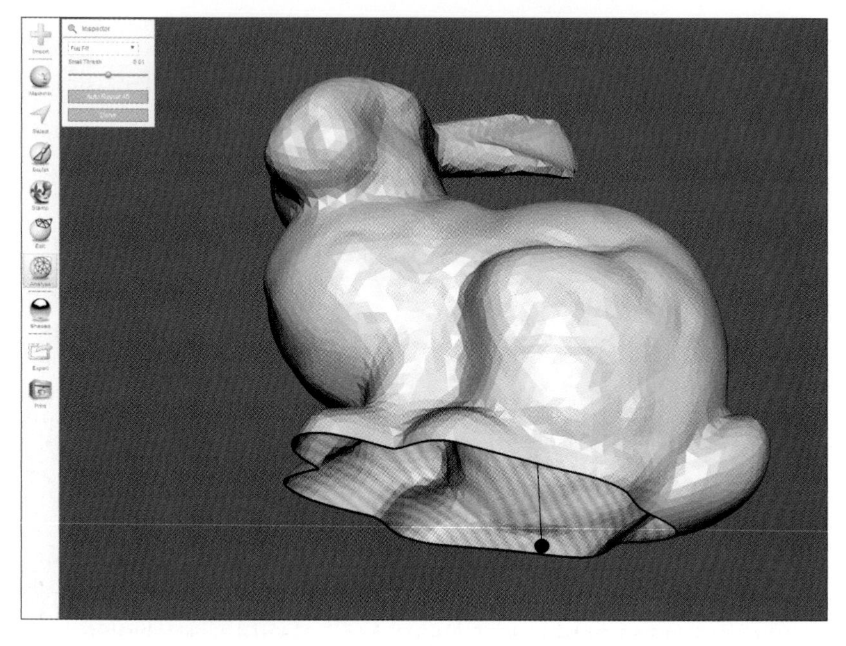

Figure 10-4. *Analysis and Inspector tools on the bunny model*

Next, click the blue pin (or the "Auto Repair All" button in the Inspector tool), and you will see the bottom of the bunny fill in.

Click Done, and your bunny will be corrected for 3D printing as shown in Figure 10-5.

Always Use Inspector

Whether you get 3D models from the Internet or create your own models, it is always a good idea to run them through the Inspector tool to see if there are any problems with it. Sometimes more advanced repair is needed, but often the Inspector tool does a great job automatically!

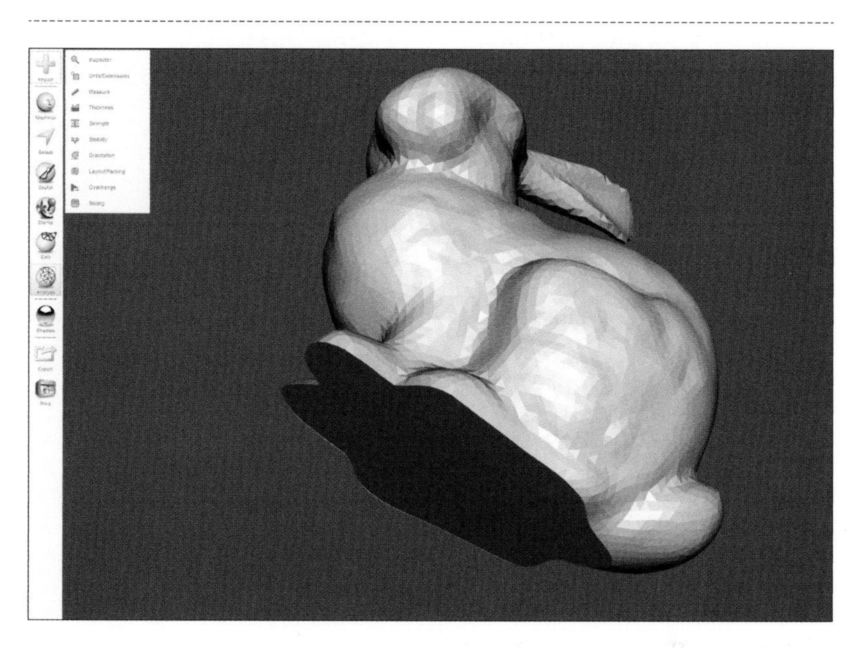

Figure 10-5. *Bunny model is fixed using the Inspector tool*

Sculpt with Digital Clay

As we mentioned, Meshmixer has more than 80 amazing tools. One of the most fun to play with is the Sculpt tool. Let's make the bunny more interesting by creating some wings to allow it to fly:

1. Click the Sculpt tool on the left menu.
2. Click Brushes and select the "Draw" tool. The Draw tool allows you to add to and create volume in the model, as if you were applying more "clay" to it.

   Before you start modifying the bunny, notice that there are settings that affect your brushes on the lefthand side. These options have a dramatic effect on how the tools work. For now, open up the settings windows to access the properties and refinement areas Figure 10-6, and adjust the settings to match the following.

   In the properties area:

   - Strength: 77
   - Size: 55
   - Depth: 0
   - "Flow" and "Volumetric" should be checked

   In the refinement area:

   - "Enable Refinement" should be checked
   - Refine: 100
   - Reduce: 100
   - Smooth: 9

3. Now, rotate your view so that you are looking at the left shoulder of the bunny. Hold your left mouse button down on the shoulder area, and while keeping the left mouse button down, start moving back and forth in a line as in Figure 10-6. If you spend a bit more time in the center than you do at the edges, you will be slowly building up a wing on the bunny.

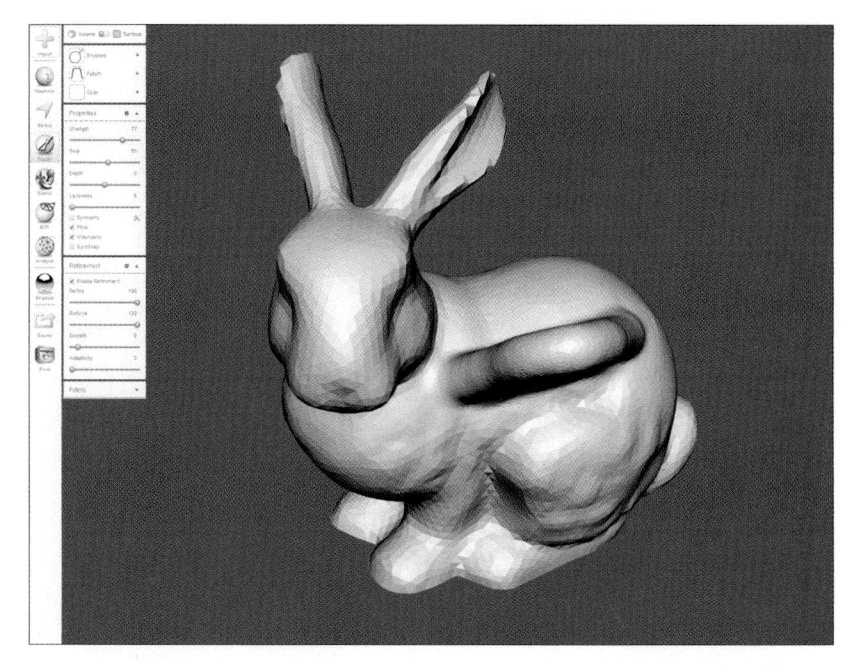

Figure 10-6. *Properties and refinement areas are adjusted in the Sculpt tool*

A Note on Sculpting

Many sculpting commands (including the Draw command) will build toward you, in other words toward the viewpoint from which you are looking at the model. So, as you are "drawing" the bunny's wing, it might seem as if you are just drawing a line... but if you rotate around, you will see that the wing is being "drawn" toward you.

4. Once you have drawn one wing, rotate your view to the other shoulder with the right mouse button, and "draw" another wing so that your bunny can properly fly as shown in Figure 10-7.

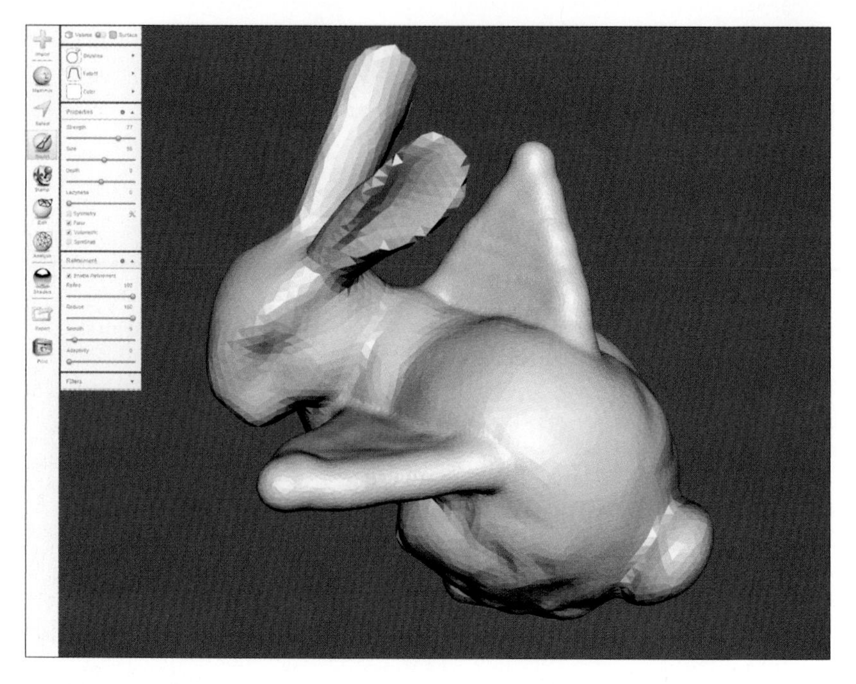

Figure 10-7. *Both wings have been added to the bunny*

Adding Support Structures

We talked about the need for support structures in previous chapters, and Meshmixer is a truly unique program in its ability to help you successfully 3D print. One standout feature of Meshmixer is the creation of support structures that will support overhanging areas of your model so that they will not droop during 3D printing. Let's create some of those support structures now:

1. On the lefthand menu, click Analysis → Overhangs.
2. Change the top option from "Custom Settings" to "Ultimaker2." You will see that certain places on your model have turned red as shown in Figure 10-8. Those are the areas that qualify for the generation of support structures, based on the settings you entered in the menu on the left. You can change these settings to what works best once you learn the capabilities of your specific 3D printer. (The Ultimaker2

printer was chosen for this example, but a different choice may be appropriate for your specific printer. You can always use one of the prefigured settings as a starting point for your own printer's needs.)

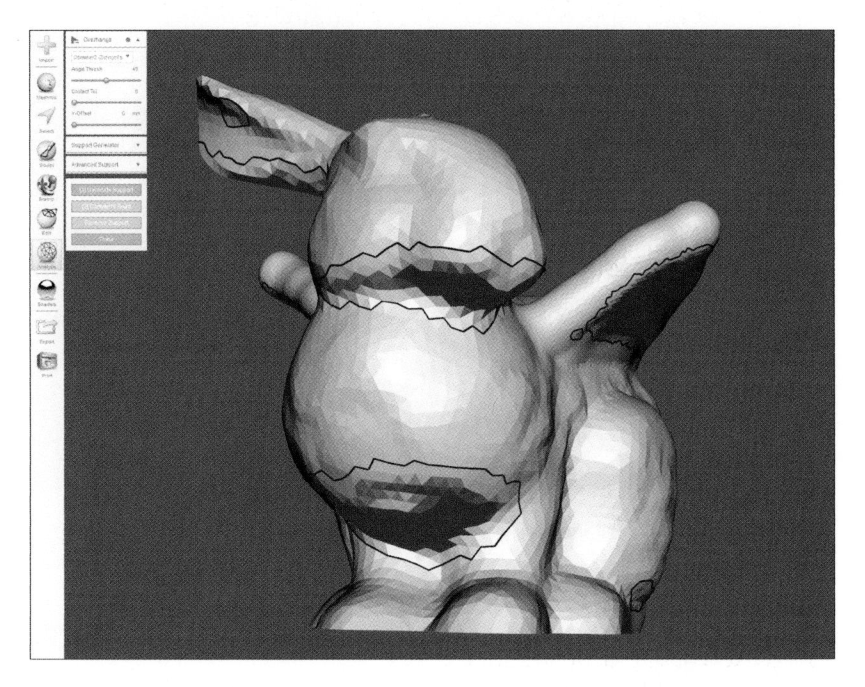

Figure 10-8. *Adding support structures to the bunny*

3. Click the "[1] Generate Support" button.

   You just told the program to create support structures to keep your model in place during the printing process. Meshmixer will think about the support structure generation for a few seconds, and then support structures will appear as shown in Figure 10-9! As we mentioned before, these support structures can be edited and removed at will, or regenerated after you change your settings. For you, learning how well your 3D printer prints at specific angles and with specific materials will be part of the learning curve.

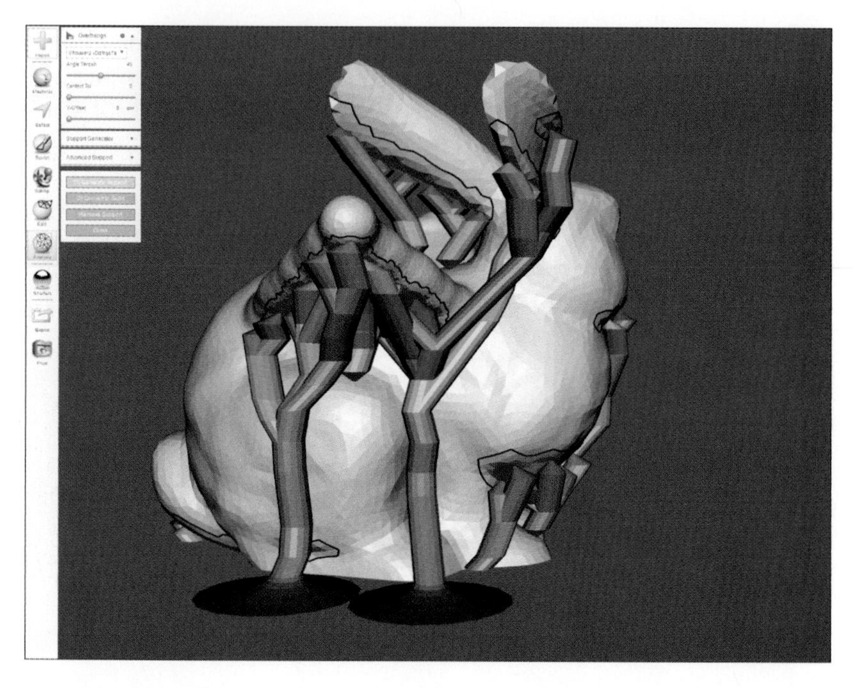

Figure 10-9. *Support structures have been added to the bunny for the 3D printing process*

4. Click "[2] Convert to Solid" to lock in the support structures as actual 3D models themselves.

5. A menu will pop up, asking if you want to create a "New Object" or "Replace Existing" object on the screen. Click Replace Existing as shown in Figure 10-10.

You will be brought out of the support structure generation utility and back into the main interface. You can click the Export button on the bottom left of the side menu to save this model for 3D printing if you desire, or you can play around with editing the bunny or even the new support structures as shown in Figure 10-11.

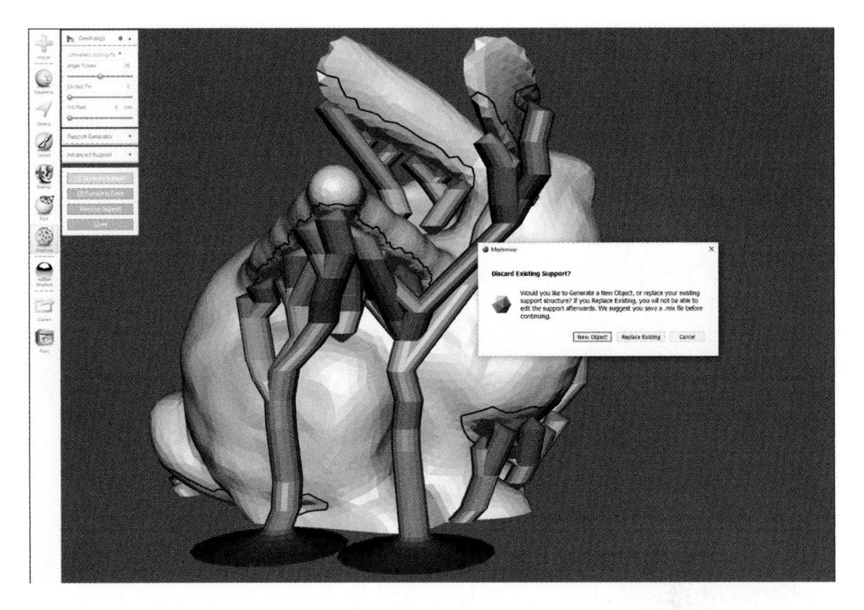

Figure 10-10. *Replacing the existing object using the pop-up menu*

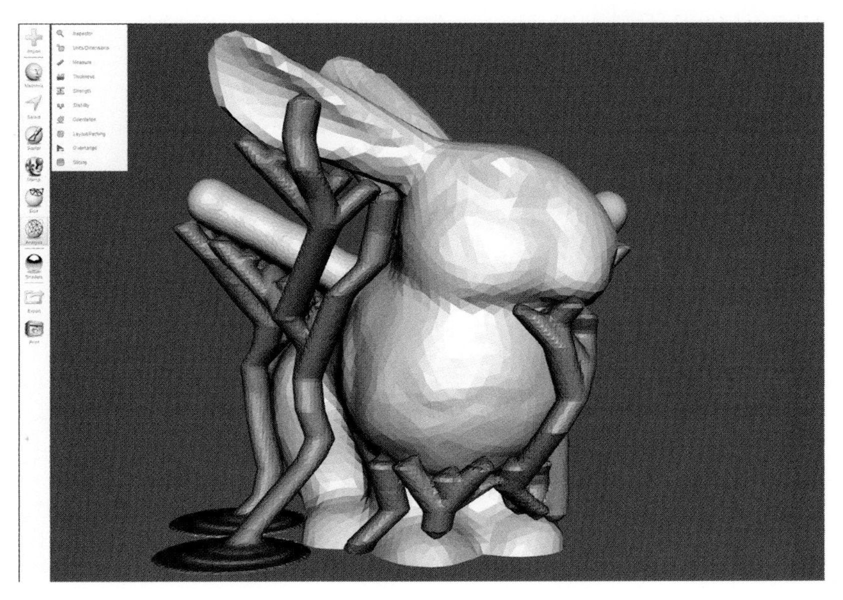

Figure 10-11. *Creating new support structures by changing the settings*

Creating a Sphere

We've shown you how to fix a model, use the Sculpt tool, and generate supports. Now let's change gears for a moment and practice on more Meshmixer tools. Let's leave the bunny by clicking the top-left File menu, and then selecting Import Sphere, as shown in Figure 10-12.

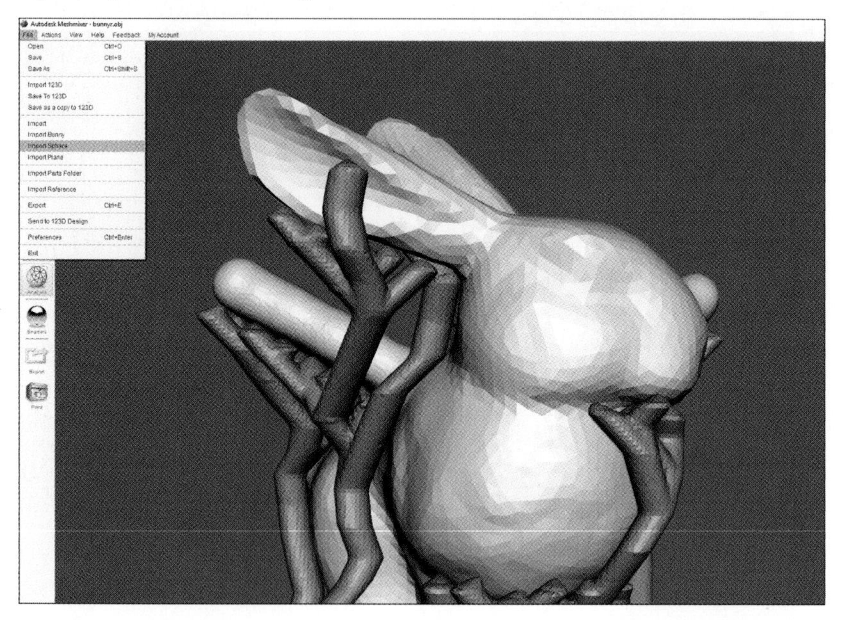

Figure 10-12. *Creating a new file by choosing Import Sphere*

Meshmixer will ask you if you want to "append" or "replace" the bunny. For this step choose "replace." You will now have a sphere, as shown in Figure 10-13.

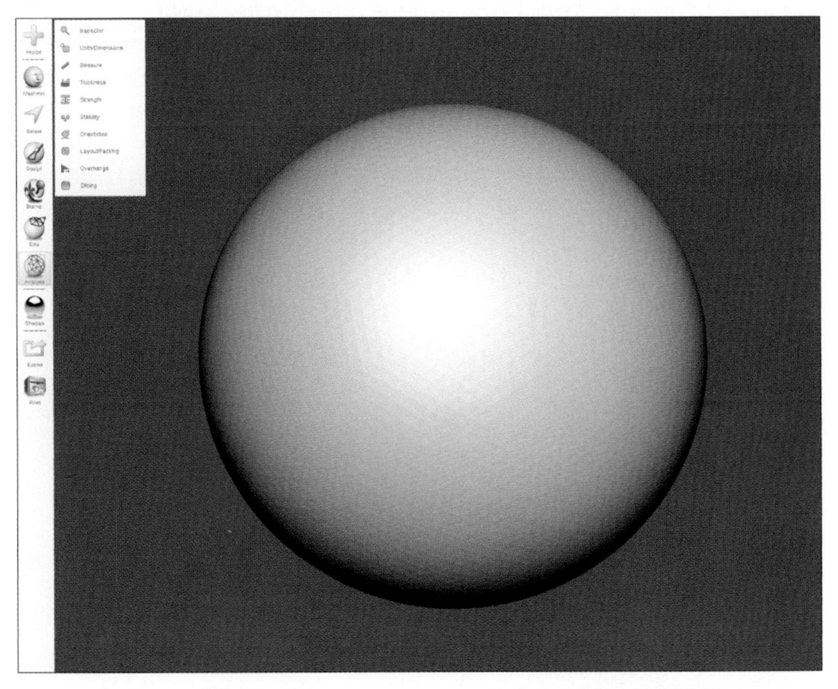

Figure 10-13. *Newly created sphere from File menu option*

You will be creating some very cool jewelry in this exercise. But we first need to start with some base geometry:

1. Start by segmenting the sphere a few times. (You'll discover why later on.) Click the Edit button on the lefthand menu and you will see a Plane Cut pop-up window, as shown in Figure 10-14.

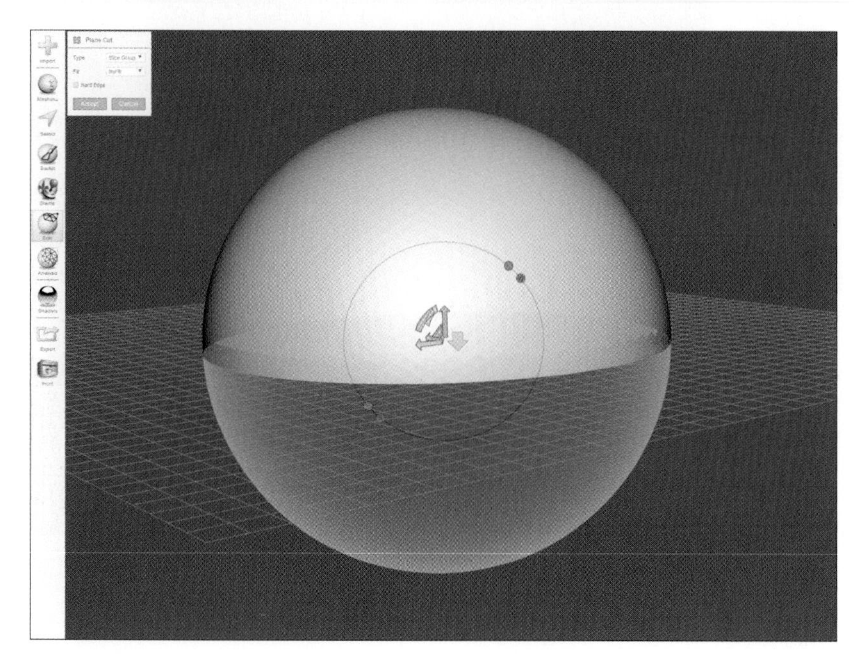

Figure 10-14. *The Plane Cut pop-up window from the Edit menu option*

2. Select the options in the pop-up box to:

   - Slice Groups
   - NoFill

3. Click Accept.

   You will see a ball that is gray on the top side and colored on the other, as shown in Figure 10-14. The "Slice Groups" command splits the sphere into two separate "FaceGroups." FaceGroups are simply areas of the mesh that allow you to manipulate them without touching the other parts of the mesh. They are denoted by a different mesh color. Your color might be different than the one shown in Figure 10-15.

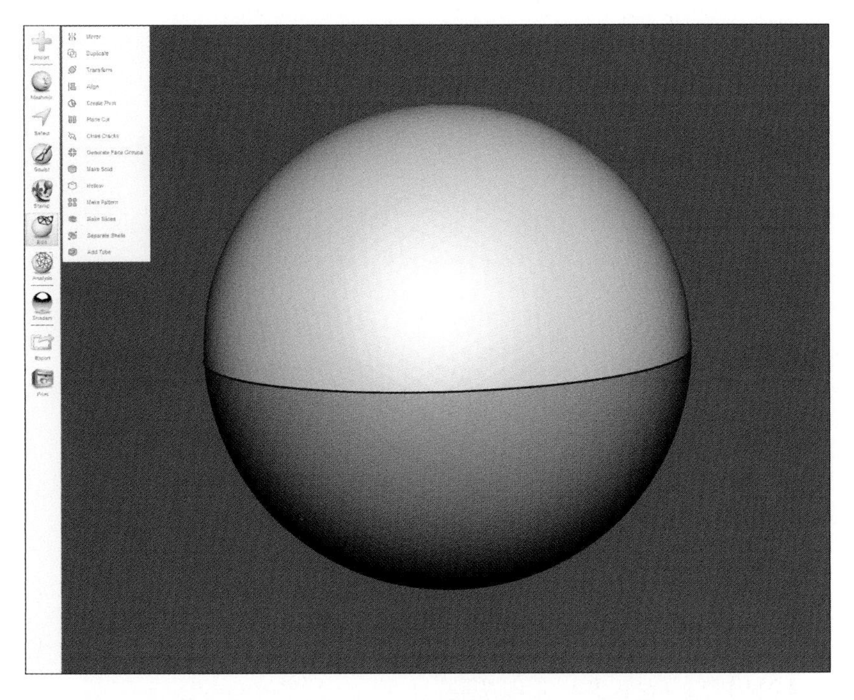

Figure 10-15. *Using the Plane Cut option on a sphere to create FaceGroups*

4. Click the Edit tool again and then Plane Cut.

5. Choose the same options as before: Slice Groups and NoFill.

   You did this before, but this time either click and drag a line with your left mouse button in another direction, or click the various arrows and bars on the "manipulator widget" in the center to move the cutting plane to a new direction as shown in Figure 10-16.

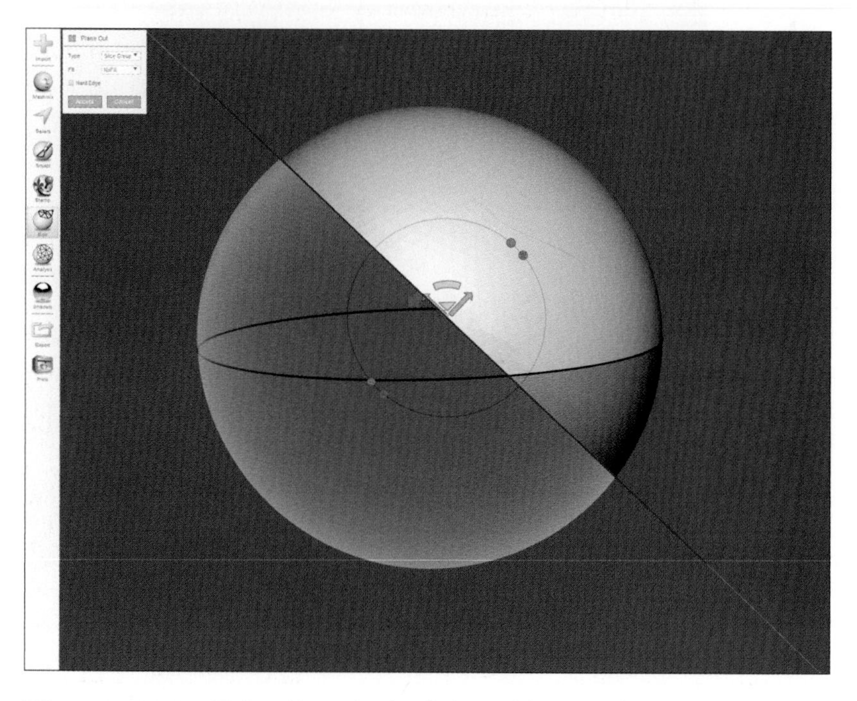

Figure 10-16. *Using the manipulator widget in the center to move the cutting plane in a new direction*

6. Click Accept and you will see a new FaceGroup appear.

   Do this procedure two or three more times, and you will get a sphere that looks like the one shown in Figure 10-17.

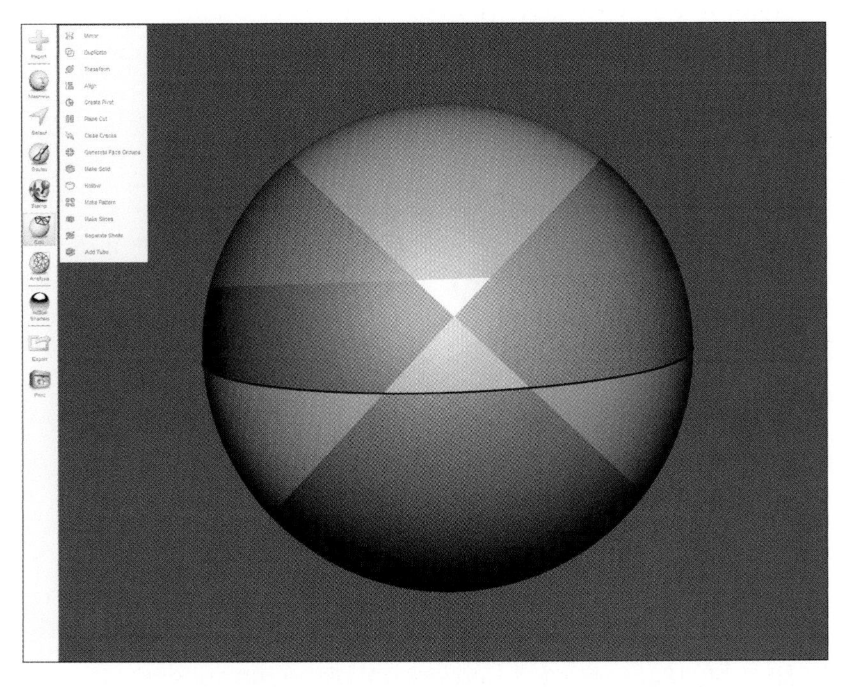

Figure 10-17. *Sphere with many FaceGroups*

Here is the fun part! "Make Pattern" is one of the flashiest and most powerful tools in Meshmixer. It is what you will use to make a piece of jewelry out of this colorful sphere.

7. Click the Edit button and go to Make Pattern.

8. Switch the drop-down menu on the lefthand side to Face-Group Borders (very top).

   You will see a preview of what will happen to your mesh, as shown in Figure 10-18.

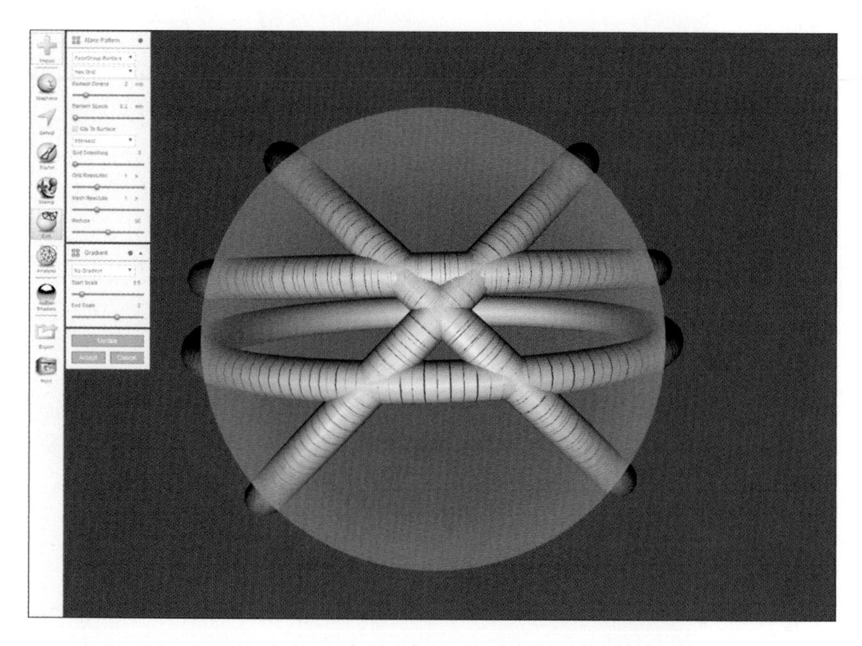

Figure 10-18. *FaceGroup Borders option on the sphere*

9. Click Accept.

   You now have a mesh that follows the borders of those Face-Groups and can be turned into a complex pendant, as shown in Figure 10-19. You could have this piece 3D printed in plastics or metals by online service bureaus. This would be considered a more advanced model to print at home, due to the amount of support material you would need, but you certainly could print it yourself.

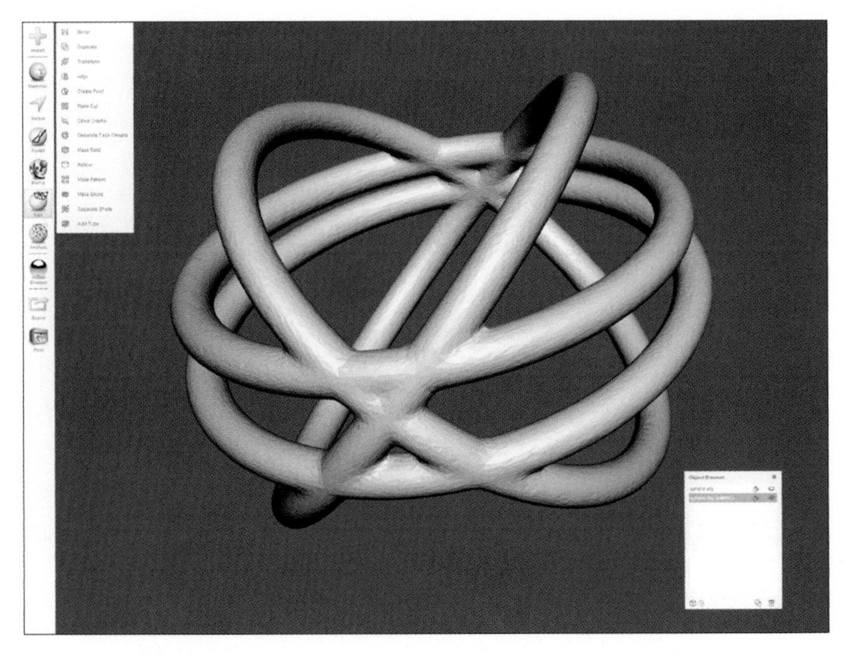

Figure 10-19. *Borders of FaceGroups made into a pendant*

Online Tutorials

If you are looking for more in-depth online training we recommend our course called 3D Printing and 3D Design Using Autodesk's Meshmixer (*http://www.honeypoint3d.com/education*). There are 159 lectures and over 16 hours of step-by-step instruction you can pace at your own level. In this course you will learn some other very popular, fun, and easy tools like "Dragging and Dropping," which is exactly like it sounds. You can pick pre-created shapes/objects and drop them onto your starter model as shown in Figure 10-20.

Figure 10-20. *Bunny shape with new arms, added colors with "Paint Vertex" tool, and Sculpt → Surface → Draw++ Stencil applied*

This has been a brief introduction to the power and versatility of Meshmixer. We invite you to play around with the other tools in this program to see what you can create. To further your understanding, be sure to also check out the online tutorials provided by Meshmixer on the Autodesk 123D YouTube channel (*https://www.youtube.com/user/meshmixer/videos*).

In the next chapter, we will introduce you to Autodesk Fusion 360, which is considered a professional-level CAD program. It is designed specifically for manufacturing but is nonetheless very affordable for individuals.

11/Getting Started with Fusion 360

While Tinkercad and Meshmixer are good, free programs, it's pretty much guaranteed that the car you drive or the computer you use were not designed with those tools. To create 3D models that can be sent to manufacturers for production, designers need to use more advanced tools such as Autodesk's Fusion 360. Figure 11-1 shows an example of the complexity of objects that can be created in this program.

Figure 11-1. *An electric race car toy that was completely designed in Fusion 360 (modeled by and courtesy of David Barrett of Virginia Tech)*

This software enables the creation of complicated models and it costs a fraction of what normal software in this league would cost. At the time of this writing, the cost of Fusion 360 is fairly unique in the software market:

- Free for personal/student use

- Free for commercial use for anyone until their business makes more than $100,000 per year
- After that, Fusion costs $25/month when you pre-pay for a year

This pricing allows users full access to a very powerful software program, without having to pay for an expensive commercial license that can cost thousands of dollars. In fact, most of the CAD models we create in our rapid prototyping business are made in Fusion 360. It is a powerful tool for professional work.

Polygonal Versus Parametric Modeling

An in-depth tutorial covering all the rich features of Fusion 360 is beyond the scope of this introductory book. We'll provide you with a basic introduction to help you understand how the program works and how to get started using it. Before we jump into the tutorial, however, we want to remind you about the difference between polygonal and parametric modeling. The STL file format that 3D scanners make, and that Meshmixer uses, generates hundreds or thousands or even millions of small triangles (which are really just three-sided polygons) to create an object. These triangles create a shell of the object rather than a solid with volume. Imagine casting a fishing net over an object. The fishing net is like an STL file that shells the object.

Parametric modeling, by contrast, uses *parameters* to define a model. Parameters are another way of saying changeable mathematical formulae. When you look at a model on the screen you are seeing how one formula interacts with another, which is why CAD programs are able to make huge changes in the model very quickly.

Fusion 360 creates what are called "solid models" that will have a defined volume and weight based on the material assigned to them. For example, you can make parametric models for objects that are made out of concrete and others for objects made out of rubber. Each of those models can have simulations run against them to illustrate what would happen to the objects in the real world if specific stresses were placed on them. After all, the density or other material properties of something in a CAD program like Fusion 360 is just another formula to calculate.

Fusion 360 is a great example of a parametric modeling program. The file output of Fusion 360 can be STL (triangles) or parametric file formats (more-than-three-sided shapes) such as STEP and IGES, which are used in traditional manufacturing.

Parametric modeling allows you to go back in time and change values based on choices you may make later. These changes can be made at any time, and they will cascade throughout the design, all the while keeping the object in proper shape. For example, if you changed the shape of a window on a house, the walls would adjust accordingly and you would not have to go back and make changes to them as well. That is because the models are based on parameters, and Fusion 360 recalculates all of the formulas accordingly.

Getting Started with Fusion 360

As we mentioned, all the features of Fusion 360 are too numerous to cover in this tutorial, however, being exposed to Fusion 360 is important if you want to someday advance past Tinkercad and Meshmixer. The following beginning tutorial in Fusion 360 is for anyone looking to add more advanced features to their CAD modeling software toolkit.

In the following steps you will be designing a ring. If you have not done so already, download and install Fusion 360 (*http://fusion360.autodesk.com*).

You will need to create a free Autodesk account to use the software. You might have already started an account if you signed up for Tinkercad or Meshmixer.

Fusion 360 is partially cloud-based. It does not require a connection to the Internet for you to do the modeling, but you *will* have the best experience on a computer that is actively connected to the Internet. The "offline" mode should only be used when your Internet connection is temporarily not available. The main computations are done on your computer, and the only time you need the Internet connection is to save and sync your

CAD design models. All files are securely saved to the Autodesk cloud for access on any computer you might be using.

When you start Fusion 360 for the first time, you will see the screen shown in Figure 11-2.

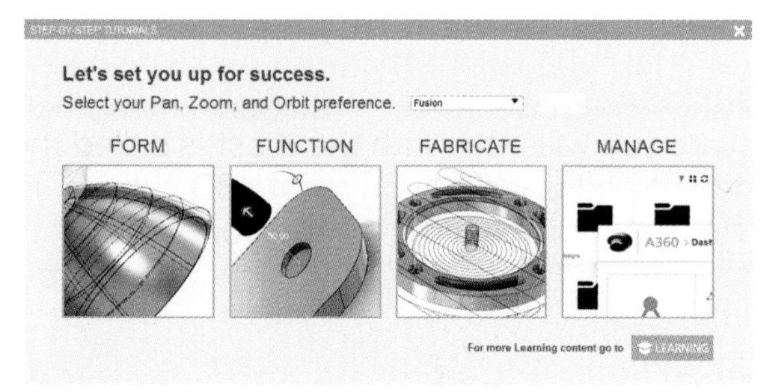

Figure 11-2. *Fusion 360 start screen*

There are four different training tutorials that show the major functions of Fusion 360. We strongly advise that you go through these at some point, if not before this tutorial. You can always access these training videos by clicking the small question mark in the upper-right part of the main Fusion 360 window and then selecting Step By Step Tutorials.

For now, close this window and return to the main Fusion 360 interface window, as shown in Figure 11-3.

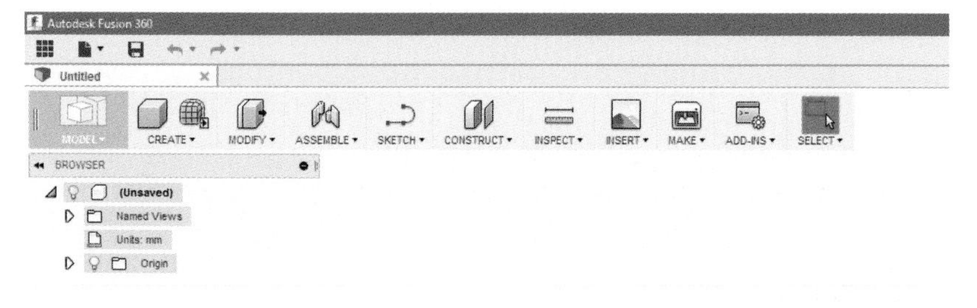

Figure 11-3. *The main interface window of Fusion 360*

Description of Interface Menu

Let's look at the different modeling environments (features) you can access in the main screen, starting from the top left and moving toward the right:

Model
: The feature you will work with the most is called the "Model" environment and is where you create 3D shapes. If you click the down arrow underneath Model you will see completely different areas of Fusion 360 that allow you to create photo-realistic renderings of your 3D models, toolpaths for CNC mills, and even a simulation workspace where you can test how your model will react to different physical stresses in the real world.

Create
: This offers two types of creation environments: one (the blue cube) for geometric shapes and the other (the purple wired box) for creating organic forms that are more fluid and able to be sculpted.

  You can have a mix of geometric and sculpted shapes in your model, and that is what we will be doing later in the chapter. If you click the small down arrow in the Create menu, you will see all the options of things you can create (boxes, pipes, sweeps, etc.).

Modify
: This feature lets you modify shapes by creating fillets (rounded edges) or chamfered (angled) corners, as well as many other operations.

Assemble
: Once you have more than one shape, you can define joints between those shapes, and see them move as a unit. This test allows you to see if there are collisions within the objects selected. This is a very useful tool before you go to the 3D printing stage.

Sketch

This menu has a lot of tools that allow you to create 2D shapes that you can later "pop out" or "extrude" into 3D shapes.

Construct

If you want to create new geometric shapes somewhere away from existing shapes, you will use this tool to construct new work planes to position your new models on.

Inspect

This feature allows you to see more information about your object. You can measure distances, or even get a cutaway view (called a section analysis) of the inside of your object.

Insert

Using this feature, if you take a reference photo with your cell phone camera, you can import that image and use it as a kind of tracing paper for your model. Also, if you have an SVG file that you imported from Inkscape or Illustrator, you can insert it directly into Fusion 360 and extrude it into a 3D shape very quickly.

Make

This function allows you to save your models as STL files, as well as giving you instant access to online 3D printing service bureaus for 3D printing price quotes.

Add-Ins

Fusion 360 also has a scripting language where you can create your own scripts to do specific things, or get "apps" from the Autodesk marketplace to enhance Fusion 360. If you know how to program, you can create specialized tools for others to use, using the add-in functionality.

Select

This feature allows you to select specific parts of the model to be modified.

Movement Box

Click and drag this feature to rotate your model in any direction.

Now that you've downloaded Fusion 360 and taken a brief tour of the features, it's time to try it out.

Making a Ring in Fusion 360

In this tutorial you will be designing a ring that could be 3D printed either at home or through an online service bureau. The steps can be followed precisely, or you can get creative with some of the sculpting functions and add your own features. It's really up to you!

Start by creating the band of the ring:

1. Click the small arrow underneath the Sketch icon.
2. Click Create Sketch. Your screen will turn a different color as shown in Figure 11-4. As you move your mouse around the screen, Fusion 360 will ask you what "plane" you want to create that new sketch on.

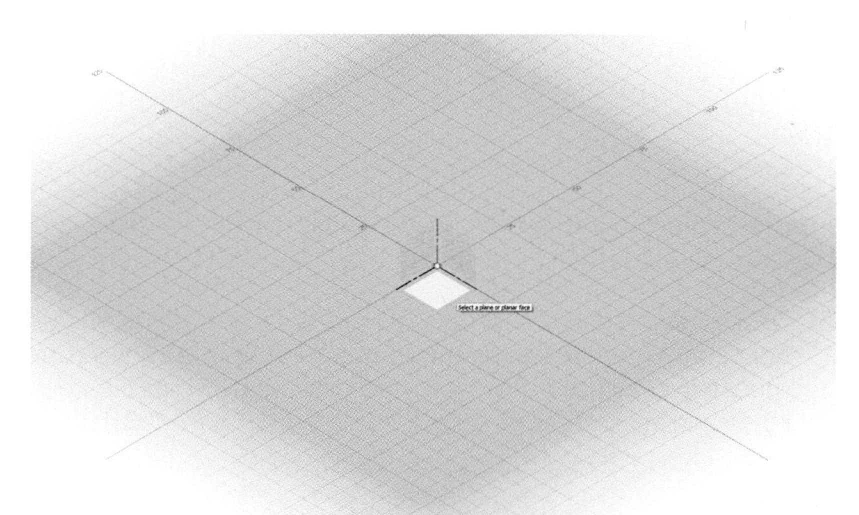

Figure 11-4. *Selected plane in Fusion 360*

3. Select the bottom plane.

   Now you are in Sketching mode.

4. Under the Sketch icon click Circle, then Center Diameter Circle as shown in Figure 11-5.

   We are creating a women's size 7 ring. For the benefit of this tutorial we researched that the inside diameter of a size 7 ring is 17.3mm.

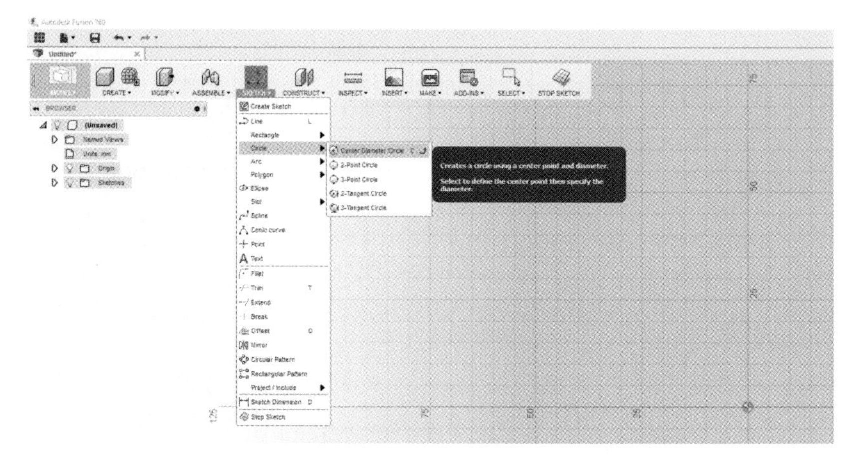

Figure 11-5. *In the Sketch mode, choose Circle and Center Diameter Circle*

5. Click once on the small circle. Where the red and green lines converge is where the origin is marked. As you move your mouse you will see how large the diameter of that circle is, as shown in Figure 11-6.

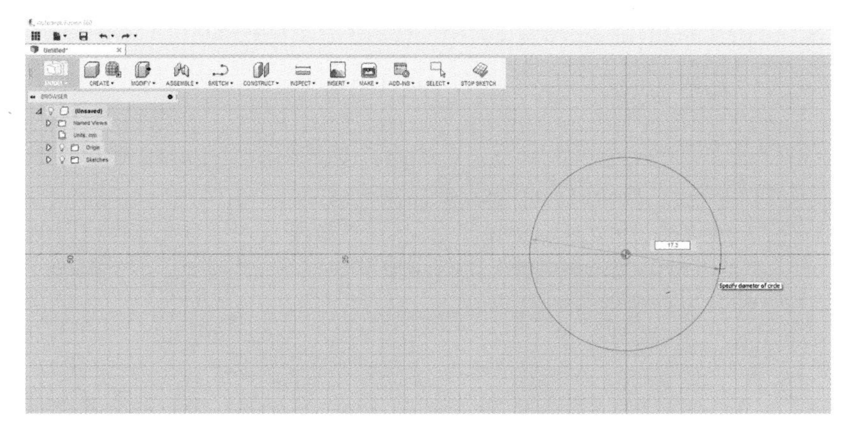

Figure 11-6. *Drawing a circle with specific dimensions*

6. Type **17.3** as shown in Figure 11-7. Then press Enter twice.

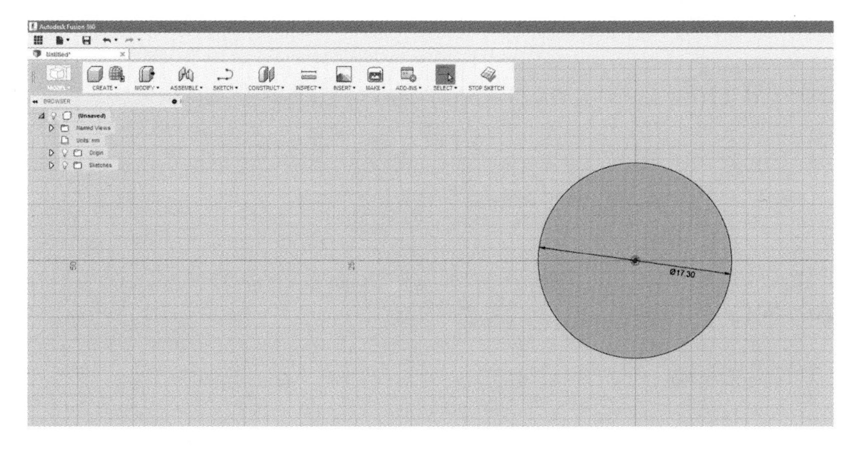

Figure 11-7. *A sketch circle with a diameter of 17.3mm*

7. That's all you need for the circle at this point so click Stop Sketch in the top right of the menu bar to get back to the main modeling environment.

   Now that we have a base circle, let's create some 3D geometry.

8. Click the small arrow under Create and select Pipe as shown in Figure 11-8.

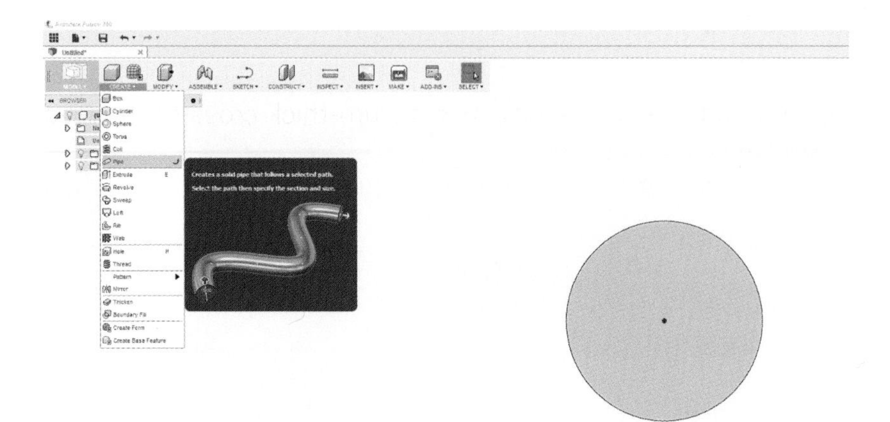

Figure 11-8. *Selection of Pipe option under the Create icon*

9. Click the edge of the circle. A 3D model will be generated and a menu will pop up as shown in Figure 11-9.

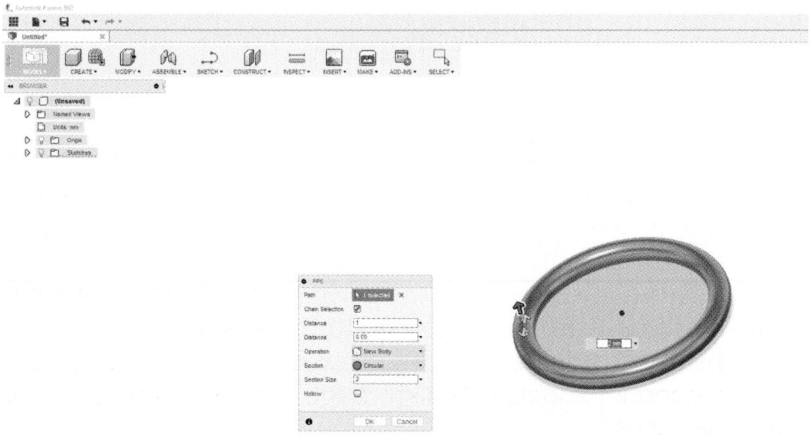

Figure 11-9. *A 3D pipe appears with a menu option*

10. Select the following settings in the pop-up menu:
    - Set both Distance settings to 1.
    - Change the Section Size to 1 and then click OK.

You just made a simple ring with a 1mm-thick cross-section, as shown in Figure 11-10.

Figure 11-10. *The completed band of the ring*

But wait!

Let's think through the measurements of this model. Remember that you created a band to fit a size 7 finger, which we know is supposed to have a 17.3mm wide *interior* diameter. But we had you create a 1mm pipe, which extends inward by .5mm and outward by .5mm, on *both sides* for a total size reduction of 1mm from what you wanted. Now the interior diameter is actually 16.3mm! If you printed this ring, it would be 1mm too small! (Always conduct a logical "walkthrough" of your steps to see if they make sense.)

Never fear, it is easy to fix these things in a parametric modeling program.

At the bottom of your screen you will find a "history" of all of the actions you took. Do you see the initial sketch there, as well as a pipe command? Let's go in and edit that sketch.

11. Double-click the first Sketch icon on the bottom left of the screen, just to the right of the Play commands, as shown in Figure 11-11. This will bring you back into the initial sketch you made.

Figure 11-11. *Arrow showing the "history" that is being built as you work in Fusion 360. Double-clicking any item will allow you to change that item.*

12. To fix the model and make it larger, double-click the "17.3" number and edit it. You can also use arithmetic operations in there as well! It is valid to type either 17.3+1 or 18.3.

13. Change 17.3 to 18.3 either by directly typing that number in, or doing the addition as shown in Figure 11-12.

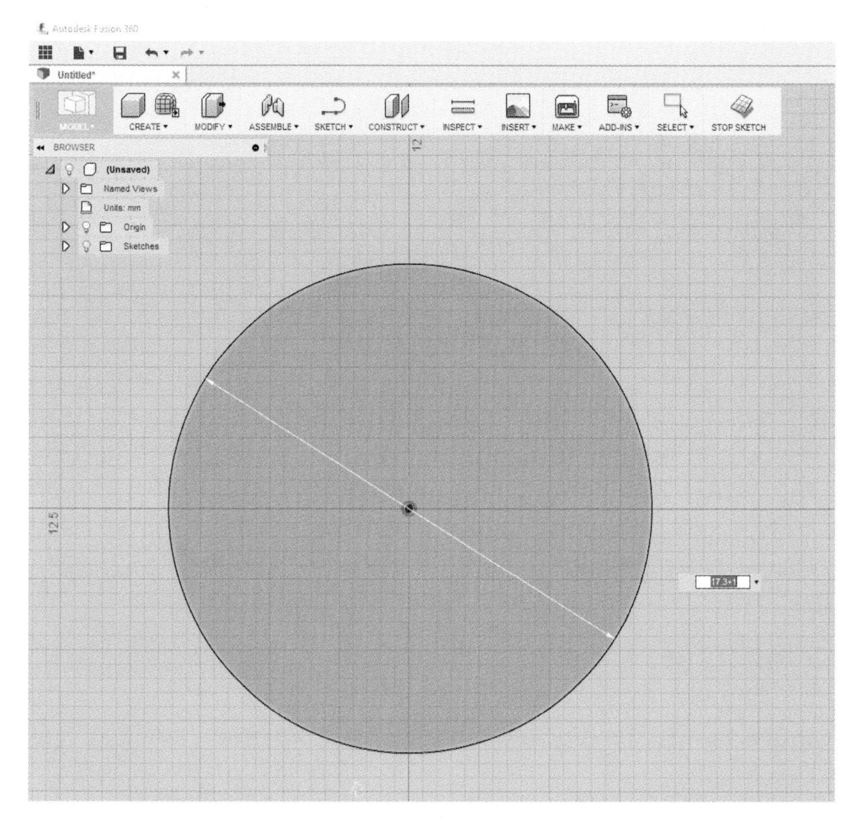

Figure 11-12. *In the sketch environment, double-clicking the diameter number allows for editing. The field entry now reads "17.3+1."*

You will now be left with a circle that is 18.3mm in diameter. Figure 11-13 shows a close-up view of the new value of the diameter.

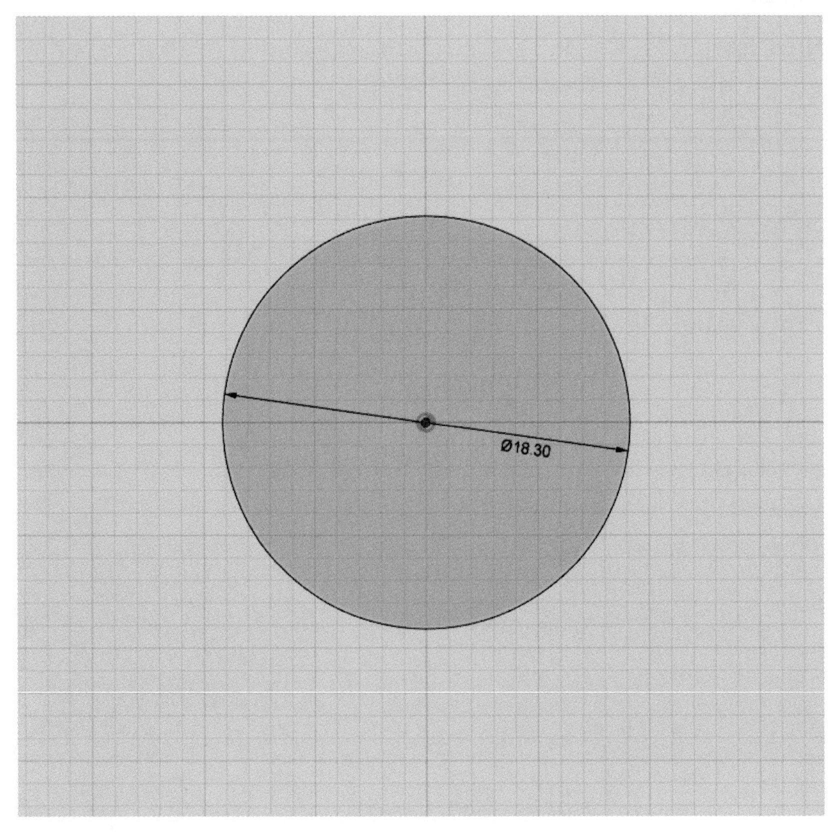

Figure 11-13. *Close-up view of changed diameter*

14. Click Stop Sketch to return to the main environment.
    Notice your ring updated in the background to 18.3mm in diameter, but the pipe is still 2mm. The effective inner size is now 17.3mm, which is what we wanted.

Adding an Embellishment to the Ring

Now, let's put a topper on that ring to make it more interesting!

15. Click the purple grid-box in the Create icon at the top to enter into Sculpting mode as shown in Figure 11-14.

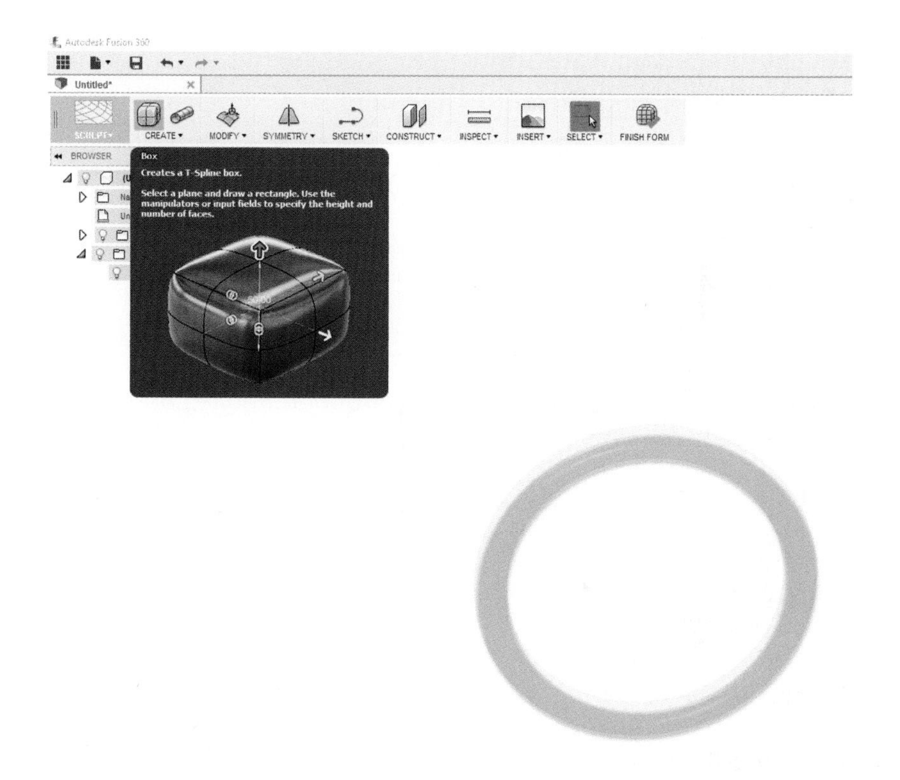

Figure 11-14. *Purple grid-box has been selected in the Create function and Fusion 360 is now in the Sculpting environment*

16. After entering into Sculpting mode, click the top leftmost shape that looks like a 4 x 4 x 4 cube with rounded edges.
17. Click the same plane as you did before to create a cube on the bottom plane. The view will reorient to that plane. Then,

position the cube around the "top" of the ring, somewhat overlapping the shaded ring, as shown in Figure 11-15.

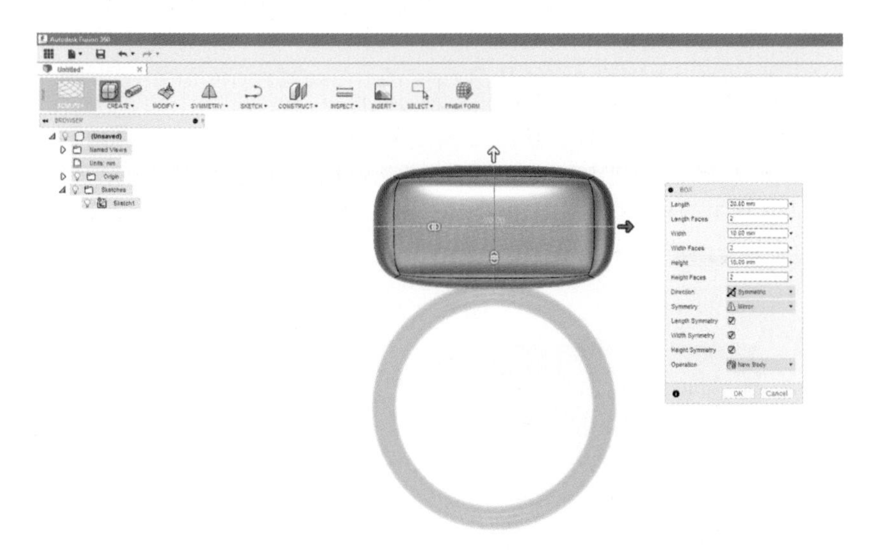

Figure 11-15. *Positioning the 4 x 4 x 4 rounded-edge cube above and touching the ring*

18. Let's pretend we are jewelry designers and happen to know the settings shown in Figure 11-15:

    - Length: 20mm
    - Length Faces: 2
    - Width: 10mm
    - Width Faces: 2
    - Height: 10mm
    - Height Faces: 2
    - Direction: Symmetric
    - Symmetry: Mirror
    - Check all the symmetry boxes: Length, Width, Height
    - Operation: New Body

    These settings are just for this project but you can play around with changing them now, or go back into your design and change them later.

Adding to the Embellishment

You just created a sculptable box of parametric clay in the last few steps, so let's make it look unique!

19. Hover your mouse on top of the box. Click one of the top faces and it will turn blue as shown in Figure 11-16.

    The other top faces will turn yellow to indicate that symmetry is turned on; a change made to one face will affect the other faces in that object.

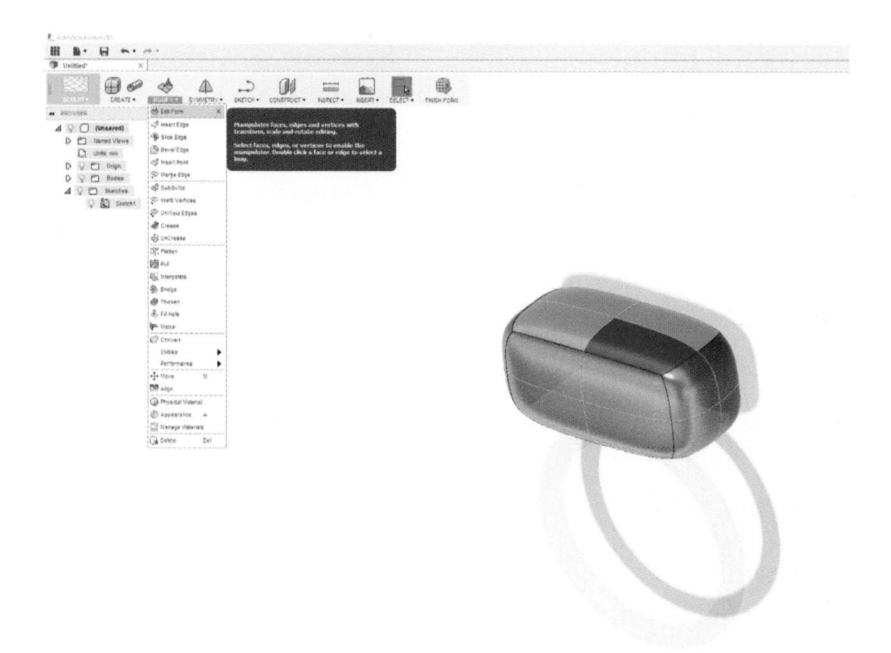

Figure 11-16. *A face of the box has been selected, turning it blue. The Modify function has been selected.*

20. On the top menu bar click the down arrow under Modify and select Edit Form as shown in Figure 11-16. A manipulator widget will appear, and as you drag those various arrows and sliders, your form will change as shown in Figure 11-17. You

might have to use the navigation cube in the top-right part of the screen by clicking it and dragging it to rotate the view.

21. Click the slider circle as shown in Figure 11-17 and move that upward a bit. This will rotate the plane inward. Because you have symmetry turned on, all of the other faces will turn inward as well.

Figure 11-17. *Twisting the faces inward using one of the manipulator widget handles*

22. Turn that slider to -45 degrees as shown in the white entry window in Figure 11-17. Don't worry about changing any other settings in the pop-up window.

    Before clicking OK, you can now manipulate other faces as well. If you happened to click OK before you were ready, that is fine; just right-click anywhere on the screen and select Edit Form again and click where you want to start sculpting.

23. Click the intersection point between all four faces on the righthand side of the cube. Drag that point inward as shown in Figure 11-18.

    You can also use Edit Form on faces, individual edges, or even the points where the edges meet. Just make sure that whatever shape you create intersects with the band of the ring that you made earlier.

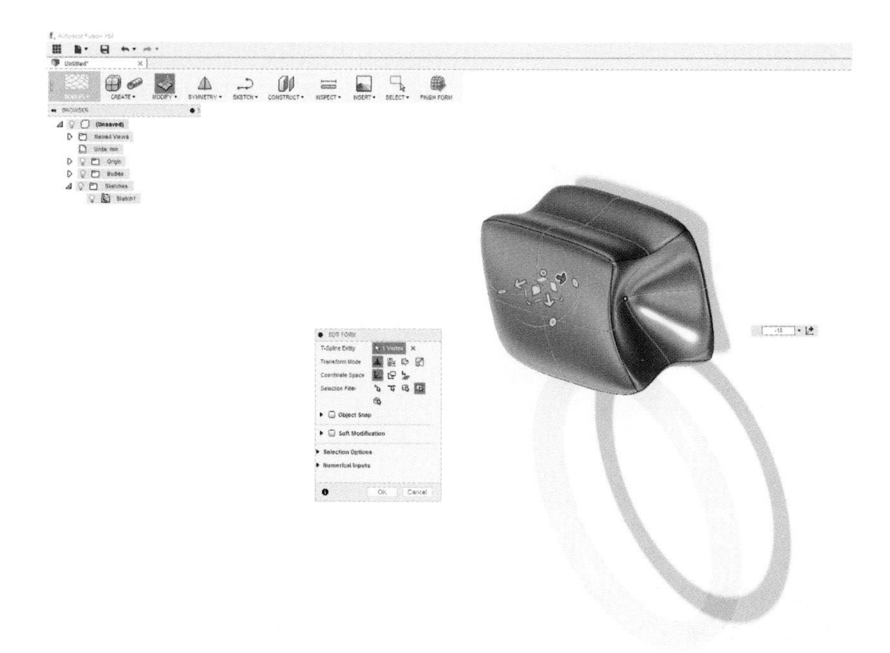

Figure 11-18. *Within the Edit Form option, a small intersection at the right was selected and dragged inward*

24. When you have a shape that you like, click Finish Form on the top menu bar to exit out of the sculpting environment. You now have a ring and a topper!

Exporting Your Model for 3D Printing

The final step is to export the model into your 3D printing software, and you're all set to go.

Fusion 360 can only export one model at a time, so you will need to combine the ring and its topper into one shape before exporting so that it all prints together.

25. Click the arrow under Modify and then select Combine as shown in Figure 11-19.

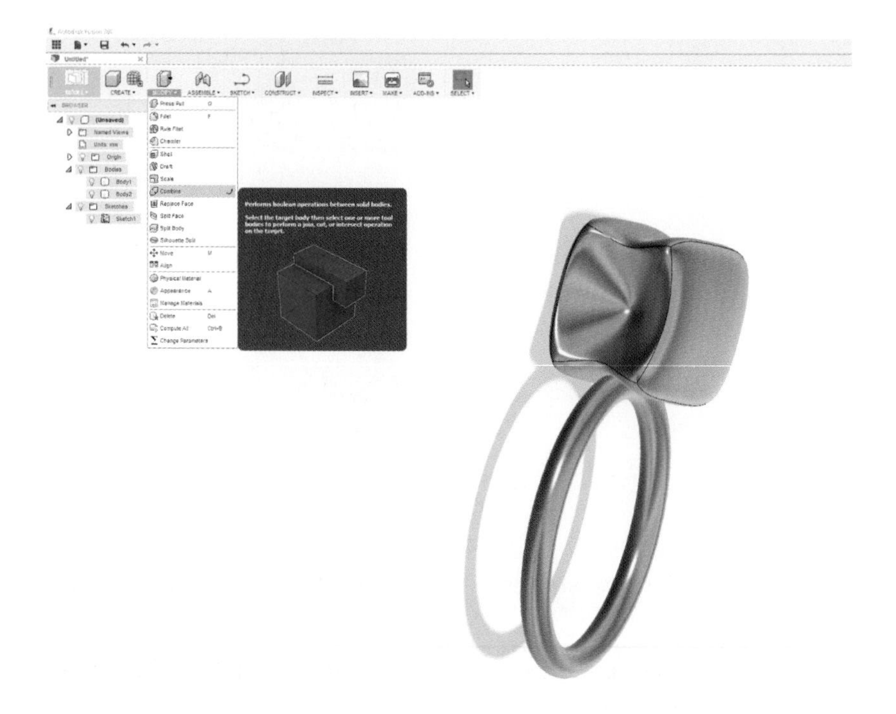

Figure 11-19. *Initiating the Modify → Combine command to merge the shapes*

26. Click the ring first and then the topper.

    One will be selected as the "Target" body and one will be the "Tool" body. Again, make sure that the Combine option is selected as shown in Figure 11-20.

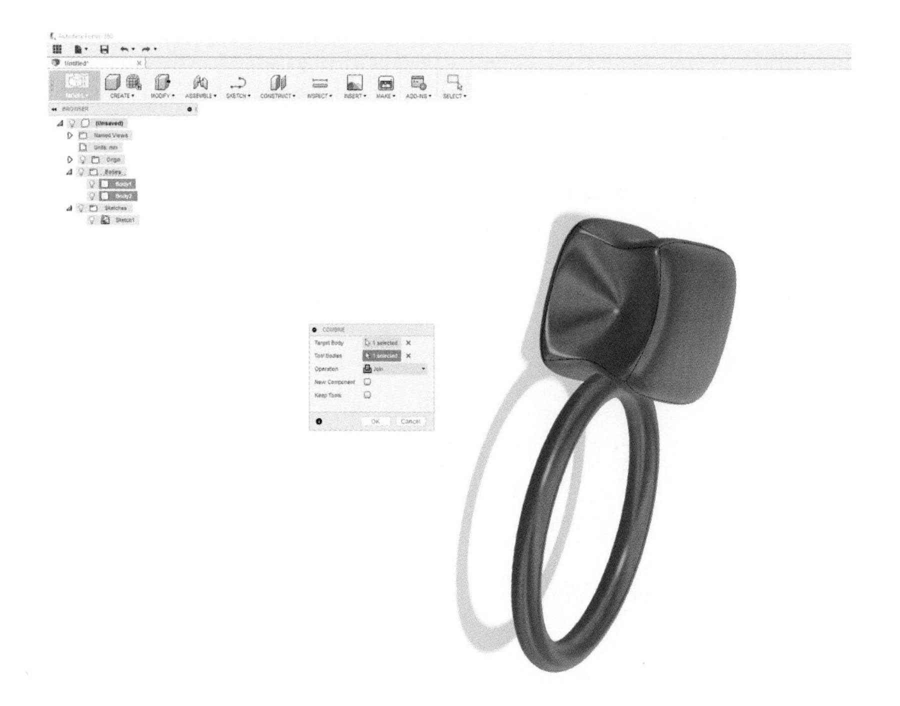

Figure 11-20. *Target and Tool bodies selected in Combine command*

27. Click OK. You now have one shape you can export for 3D printing!

28. Under the "Bodies" hierarchy in the left menu, find the single body you just created, and right-click it.

29. Select the "Save As STL" option and select "High quality" as shown in Figure 11-21.

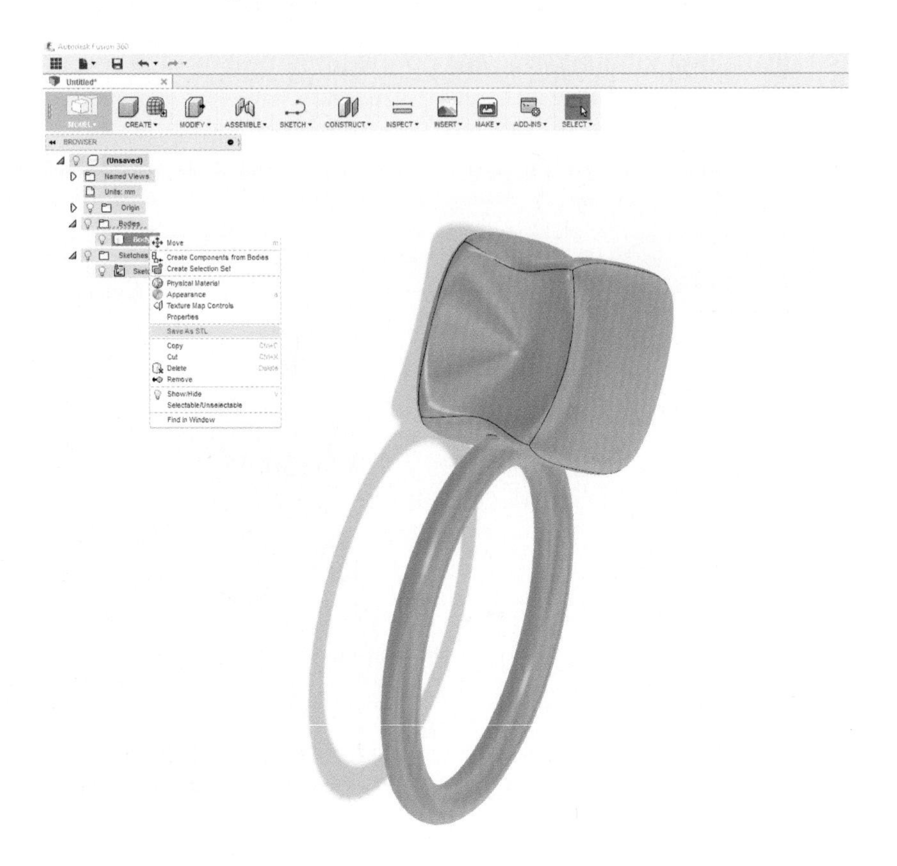

Figure 11-21. *Selecting Save As STL on the combined body*

Congratulations! You are all set to load the model into your 3D printer's slicer software, generate support structures, and print a 3D object.

Combining Shapes from Meshmixer

Remember Meshmixer from the previous chapter? You can bring the ring model into Meshmixer by importing the STL file, and then you can add organic shapes as shown in Figure 11-22.

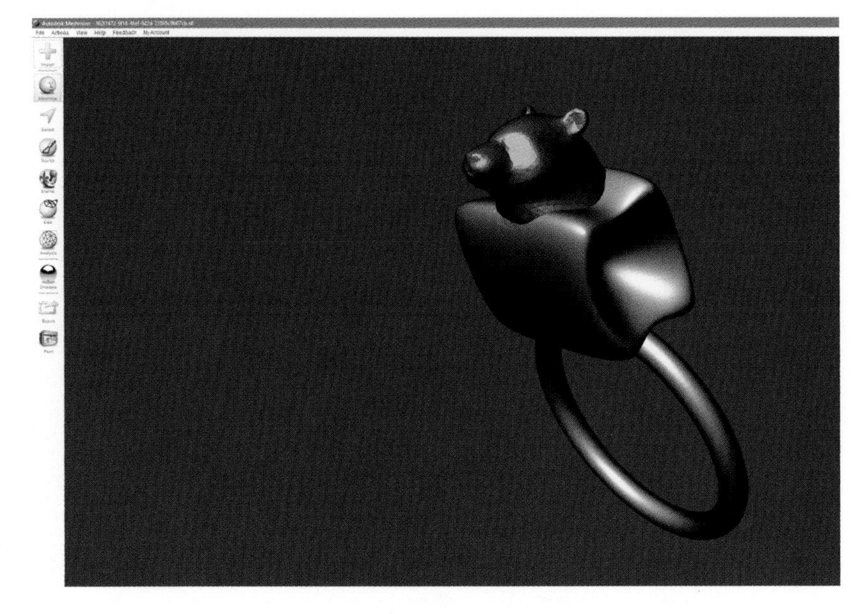

Figure 11-22. *Your ring with topper imported into Meshmixer with a premade bear's head shape added*

In this chapter you learned how to create a ring with a specific diameter, attach a custom sculpted topper, and combine them together to make a finished model that is ready to export for 3D printing. If you created a version of your own that you really like, post it on Thingiverse.com (or another online 3D model repository) and share it with the world!

In the next chapter, we'll talk about setting up your own makerspace for 3D printing. It has practical tips on what to buy, use, and display to have the best 3D printing experience!

12/Setting Up Your Personal Makerspace for 3D Printing

Whether you call it a makerspace or FabLab or hackerspace or community workshop, the names all refer to community spaces where creativity and exploration are valued over "getting it right the first time." These community-operated physical spaces are where regular people with common interests can meet, socialize, and make.

The people who use makerspaces share one thing first and foremost: the desire to experiment, tinker, invent, and learn. The actual tools that exist in a makerspace come second to that inventive drive, and in fact, many makerspaces start out with nothing more than a group of people coming together to help each other learn. Often makerspaces will have 3D printers, laser cutters, CNC machines, electronics components, and various hand tools, as shown in Figure 12-1.

Figure 12-1. *FabLab: The Science Dissemination Unit (SDU) in Trieste, Italy (Moreno Soppelsa/Shutterstock.com)*

The goal of this chapter is to help you create much the same feeling and functionality in your home as you would find at a public makerspace. Creating the perfect makerspace is not about how many tools and pieces of equipment you have. Rather, it is about making sure you are able to be creative and productive.

So, don't feel as if you need to spend a lot of money on fancy tools or have the perfectly organized space (Figure 12-2) in order to make a space you can create in!

Figure 12-2. *The perfectly organized mythical makerspace*

In reality, your makerspace likely won't be as neat as in the image. Even with the best of intentions, your makerspace will more likely end up looking like a mad scientist lives there (Figure 12-3) and that's actually a good sign!

Figure 12-3. *What a real 3D printing maker's workstation looks like (Moreno Soppelsa/Shutterstock.com)*

Getting Ready

Here are three essential tasks we recommend you do to get ready to start 3D printing at home, along with advice on how best to accomplish them. You'll find it's really helpful to have your space well set up before you begin! We've provided a checklist at the end of the chapter so you can make sure you have the essential tools.

Task 1: Research Which 3D Printer to Buy

You don't need a 3D printer to participate in this technology but a 3D printing space wouldn't be the same without... well... a 3D printer. In Chapters 4, 5, and 6 of this book, we identified 3D printer qualities to look for when purchasing a 3D printer, but this section will give you more insights.

At last count, there were more than 200+ 3D printer manufacturers in the consumer market. They range from startups to multimillion-dollar companies. To get a real sense of the manufacturer's quality, we recommend you visit their forums and read the comments other owners have made.

Decide how much or how little time you want to spend building and maintaining the printer yourself. We put together our first 3D printer, the Mendel Max 2.0, as shown in Figure 12-4, on our dining room table. It took 16 hours over the course of 3 days. We learned a lot from the experience, but, if you don't have the time or patience, we recommend you buy one already assembled. Pre-assembled 3D printers are more expensive but worth the money if you would rather focus on 3D CAD modeling or 3D printing rather than learning all the "ins and outs" of how a 3D printer works. Assembled or not, you will still have to learn how to maintain the 3D printer. For tips, read Chapters 4, 5, and 6.

Figure 12-4. *Mendel Max 2.0 assembled kit (makerstoolworks.com)*

You may also want to take a look at *Make:* magazine's annual issues devoted to 3D printers or visit their companion website (*http://www.makezine.com*). The magazine's editors select dozens of the latest 3D printers to rigorously test and score.

Task 2: Create a Safe Work Area

Whether you decide you want your makerspace to be in your garage, home office, or shed, you will want to make sure you have a safe environment and one that is conducive to your well-being.

You want to be able to move around the workspace easily, have good lighting, tape down your various power cords so you don't trip over them, and have a fire extinguisher within 6 feet. Just in case.

In general, 3D printers are safe machines, but you will want to take precautions to make them safe for other members of your household, including spouses, kids, pets, visiting relatives, and curious neighbors. Certain components of the 3D printer (like any machine) need to have some "look only" areas explained in order to keep everyone safe.

One example of a "look only" area is the extruder (or hot end). You should take caution to not allow anyone to touch the extruder assembly while it's in operation, as shown in Figure 12-5. The nozzle can reach hundreds of degrees in temperature.

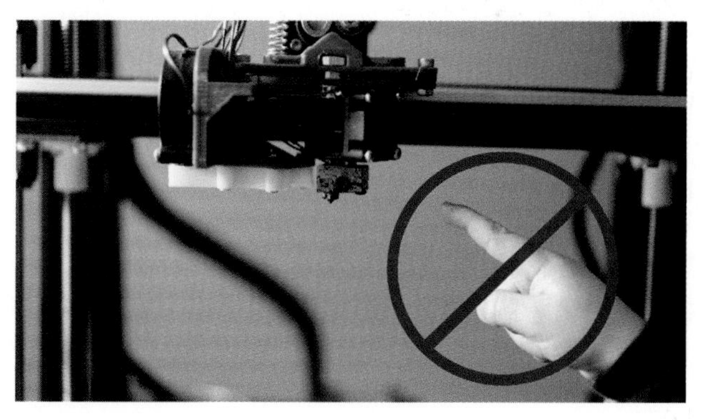

Figure 12-5. *Some 3D printers have easily accessible heating elements. Avoid touching the extruder assembly when the 3D printer is in operation. (We turned it off to take this picture!)*

There are other components you need to watch. Many 3D printers have belts and rods that move, which can present pinch hazards. Exercise caution and make sure to keep fingers, hair, and clothing away from the moving parts when they are in operation (Figure 12-6).

Figure 12-6. *3D printer belt and pulley crushing paper, showing the potential pinch hazards*

Many 3D printer kits have exposed electrical wires and non-enclosed electrical components. 3D printers also run on electrical power from a wall outlet, so normal appliance caution is warranted.

When you are removing a print from the print bed with a sharp spatula, watch your fingers to avoid getting cut. When removing support material from a 3D print, wear protective glasses and make sure your peripheral tools (picks, rotary sanding tools) are stored in a safe place. There are more safety tips you will learn from user forums, the manufacturer's website, and experience.

Good Ventilation Is Recommended

Although many say that the fumes from heating filament in a 3D printer are safe, we feel it's prudent to run a HEPA filter near your 3D printer and/or to have some sort of ventilation near an open window that isn't blowing directly on the 3D printer. The filament material ABS, for example, smells much stronger than PLA for a reason. The "S" in ABS stands for "styrene," which is a known irritant to the respiratory system.

If you want to learn more, take a look at a study done by the University of Illinois (*http://bit.ly/1DSjOd0*) that tested 3D printers for fumes emitted by FDM printers. The study found that 3D printers while in operation produced nanoparticle emissions similar to or higher than cooking with a gas stove. These nanoparticles can be absorbed more easily into the respiratory system than other types of particles so it is important to provide for the best ventilation you can. The study did not say what effects those nanoparticles would have on a typical person, but the study did state that ABS produced 10x the emissions of PLA—which is why we recommend printing with PLA.

Task 3: Choose the Best Physical Environment for Your 3D Printer

One of the nice things about creating a makerspace for your 3D printing work area is that the actual equipment, tools, and accessories can be contained in a relatively small area. Many people even use 3D printers directly on the same desk as their computer! We recommend, however, that you use a table that is dedicated to the 3D printer, and we'll tell you why a little further down this section.

In the beginning of our business, we had more failed 3D prints than we care to remember. It often took hours of trial and error to figure out what factors actually caused the failure. You can get a jump ahead by learning from our experience! Here are our top items to consider when choosing the best environment for your printer:

- Does the room temperature fluctuate too much? Garages are notorious for heat and cold fluctuations. In the summer they are too hot, in the winter too cold. Too-high temperatures can negatively affect the viscosity of resin, as well as causing stepper motors to not work properly and potentially overheat. Find a place where you can control the temperature and keep it around 70 degrees Fahrenheit. We print in our garage, so we bought a powerful heating and cooling air conditioning unit to moderate the temperatures. Check the BTU ratings before you buy, and make sure the ratings match the size of your room.
- Do you have kids? We have a very curious three-year-old who loves to get his hands on anything that is off limits. Assume little hands will want to touch the sound-making, toy-producing machine and place your 3D printer appropriately out of their hands' reach. In addition to their safety, any stray touch can upset the calibration of the printer bed and cause prints to fail.
- Beware of airflow and sunlight. An errant breeze from an air conditioning unit that you cannot feel unless your hand is behind the printer can spell disaster on your printing process, cooling some sections of your print but not others. An open window can bring some unwanted heat or air currents as well (you *should* have proper ventilation; just be sure to place the printer not in line with air currents). If you're printing with resin, you will want your print station away from sources of UV light in order to not prematurely cure your resin!
- Beware of unwanted movement. Whether you have a high-traffic area or an unstable table that jerks too much when the 3D printer head is moving back and forth, realize that every bump and shake of the table can affect your printer in unwanted ways! Earlier we mentioned that it's not a good idea to have your 3D printer on the same desk as your computer. This is why.

- Your workstation should be within good distance of a power outlet that you can use with surge protectors. It's preferred that your work table not be placed against a wall so that you can move around all sides freely. 3D printers need maintenance, and some have components located on the back that you might need to access frequently.
- 3D printers make noise. Some models can take hours or days to print. We recommend not placing the printer in your bedroom, of course, or near a shared wall of your bedroom.
- If your printer does not come with the ability to self-manage its own prints, you will need to dedicate a computer to the printing process for the entirety of the print. Make sure your computer does not enter into sleep mode during the print!
- Store your materials properly! We talked about proper storage and handling of both filament and resin in Chapters 4 and 5 respectively, but it is worth stating again. Many filaments absorb moisture from the air, and keeping that moisture out will be one key to successful prints. Though resin does not have moisture issues, temperature changes can cause failed prints when working with resin.

Setting up shop in your garage? Here are some special considerations and suggestions to keep in mind:

- Dust is not your friend. Keeping the garage area, floor, and workspace free of dust will help prevent dust from being an unwanted part of your print!
- As mentioned before, try to get the temperature to stay as consistent as possible. We have a lot of windows in our garage, so using 3M foil tape helped keep the temperature inside more consistent.
- Keep open windows about 6 feet away from your printer to allow for proper ventilation. Or use a HEPA filter that is rated for VOCs. Weather permitting, open the garage door to get good ventilation.
- If possible, the environment in your garage should be relatively dry because too much moisture could render the filament unusable. Filament should remain dry and stored

properly in a plastic bin with silica packs, on the floor of the garage.

- If you keep a refrigerator or freezer in your garage, set up your workspace on the opposite side of the garage. We don't recommend that your printer share power outlets with other major appliances like refrigerators, freezers, washing machines and dryers, etc.
- Learn more tips from online sources like the Google+ community for 3D printing (*http://bit.ly/1TUFcrx*). This community has over 200K members and is very active.

Things We Recommend for Your Home Setup

Whether you are printing with an SLA or an FDM printer, your physical home makerspace setup will be similar. Keep in mind, though, that SLA printing requires an extra water source/post-processing area. Following is a list of things we recommend you have for your personal 3D printing setup. As you develop your skill and techniques, you may add more items to your list.

Setting up the space:

- Have at least a 3' x 3' x 3' space for the printer and your work area.
- Get a standard tabletop or workbench. You may need one or two of these depending on the size of your 3D printer. Each should be at least 24" x 24" with a 28" sitting height.
- Locate and use a nearby electrical outlet with an attached surge protector to protect the printer.
- Allocate a space for your computer (if your printer requires a physical connection to one).
- Have handy a standard USB cable to connect the printer to your desktop computer or laptop (if necessary). Your printer may have come with one.

Printing and processing items:

- Have one of the following to help prints stick to the build plate: extra-strength glue stick, blue painter's tape (try to find 2" wide or wider if possible), or for those of you printing with a heated build plate, Kapton tape. If using painter's tape, avoid brands with excessively waxy residue that might prevent adhesion.
- Get a 3D printer toolkit (for edging and scraping excess glue off the print bed after printing). Octave 3D Printer Tool Kit A is a good choice for around $20.
- Buy a thin spatula (to help remove prints from the build plate). We like using a thin cookie spatula, sometimes marketed as a scrapbooking tool in arts & crafts stores.
- Consider a small blowtorch for resurfacing and finishing print jobs.
- Buy flush-cut wire cutters. We recommend the Hakko CHP-170 Micro Clean Cutter with 16-gauge maximum cutting capacity (usually available from Amazon and other online retailers for under ten dollars).
- Buy wide angle and needle nose pliers (for support structure removal, as well as for removing parts from the build plate).

Cleaning up afterwards:

- A small wastebasket for discarded pieces of the print or small pieces of filament you cut off when changing filament. Or, for a failed print.
- Paper towels (and water) to clean the build plate if using the glue stick.
- Clean cloth to remove excess glue from the print bed after completing a print.
- A wire brush to clean the extruder's toothed gear if the teeth get jammed with filament, or if the nozzle gets small pieces of debris on it.
- .3mm (.012") guitar string to feed back up through your nozzle if you get a clog.

Materials and storage:

- An airtight bin you can store your filament in to control moisture absorption. Five-gallon paint buckets (with lids) are a great option here!
- Rechargeable desiccant canisters to keep the air in the bin as dry as possible.
- A suitable supply of "feedstock." This would be filament for the FDM printers or resin for the SLA printers. For FDM, a starting purchase of just one spool will last quite a long time, and somewhere around 1 liter of resin will be a good starting amount.
- Large storage unit (preferably with wheels) to store materials such as unused filament, tools, cables, acetone, etc.

Miscellaneous:

- Good lighting (via light fixtures or natural, but indirect, light). Remember that direct light from a window can negatively affect the ambient room temperature.
- Indoor fans/heater (if the room doesn't have a way of regulating the temperature already).
- Zip ties for securing wiring when repairing or tinkering with the printer parts. (But hey, you are a maker and makers get creative with the tools they need. If you have string or wire tie-backs... use them.)
- Good WiFi access in your makerspace, so you can easily access online tutorials and help on your laptop, desktop computer, tablet, or smartphone, as needed.

Additional items for resin printing:

- Protective nonporous gloves (latex or nitrile). Either disposable or not, but the resin does not really ever dry, so those gloves will need to be kept someplace where they will not get resin on other things if you use nondisposable ones.
- A printing area away from direct sunlight, with good ventilation to help mitigate the chemical smell of many resins.

- A printing work area with an ambient temperature that is around 70°F / 21°C to prevent resin from becoming too cold or too hot (affects viscosity).
- A chisel and hammer to get tightly bonded prints off of the print bed.
- Plastic container to wash your print off in isopropyl alcohol in order to remove uncured resin from the surface. This could be as simple as a plastic bag (running the risk of crushing delicate prints, though), or a small plastic food container that will be large enough for the maximum size of your printer's build area.
- 90% or above (the higher the better) isopropyl alcohol for cleaning the prints. Denatured alcohol also works, but isopropyl alcohol is easier to find and relatively easier to work with.
- Natural sunlight to post-cure the prints.
- In the absence of sunlight a container with some UV lights installed. The UV lights need to be of the proper wavelength to cure the specific resin you are using (see the resin curing instructions from your resin manufacturer to find this out).
- One extra resin container for every type of resin you print with. This is used to store the used resin so you do not contaminate the new resin with the used resin.
- A small strainer to strain the used resin from the build process, mistakenly cured resin, as well as hair, dirt, etc.
- A supply of extra build vats/resin containers. For the majority of resin printers, the resin vats are considered consumable, and have a finite lifespan, so having extra ones is a good thing.

We realize that this is a long list of "things you should consider having," but you probably own many of the items on this list already, and many of the things you probably don't have are commonly available (see Figure 12-7). In the beginning, get only what you absolutely need and purchase new supplies and equipment as you go. It may seem like a lot of prep work, but creating a space in your home dedicated to creativity and exploration can help provide a sense of purpose and fulfillment and is worth the investment.

Figure 12-7. *A loosely organized set of home workspace tools (multimeter, rotary tools, eye protection, soldering iron, and more)*

In the final chapter we will explore how 3D printing will change your life by affecting your daily living, job, and access to technology. Let's explore where this field is heading!

PART IV

The Future

13/How 3D Printing Will Change Your (and Everyone Else's) Life

In this book, you have learned how 3D printers work, how to set up and use your own 3D printer, and how to create a personal makerspace for 3D printing. You have also learned about some of the third-party services that are available to handle aspects of the 3D printing workflow for you. But the learning doesn't stop here. You are an early adopter of this emerging technology, and it will continue to grow and change rapidly in the coming months and years.

For that reason, we urge you to engage with the 3D printing community. There are many member-driven groups you can join. One of the largest online groups for 3D printing is the Google+ community for 3D printing (*http://bit.ly/1TUFcrx*), with over 200K members. Figure 13-1 is a screen capture from the website. We recommend that you find an online forum that suits your level and interests, upload your 3D model creations to online repositories, and attend 3D printing events in your community. It's a great way to stay informed and up-to-date.

We don't have a crystal ball to see the future, but we do know that 3D printing will increase in the coming years as more content is created, standards are defined, and the process of creating 3D models is made easier. 3D printing will change how we make everything from pharmaceutical pills to hardware tools. Here are some examples of what we think you can expect.

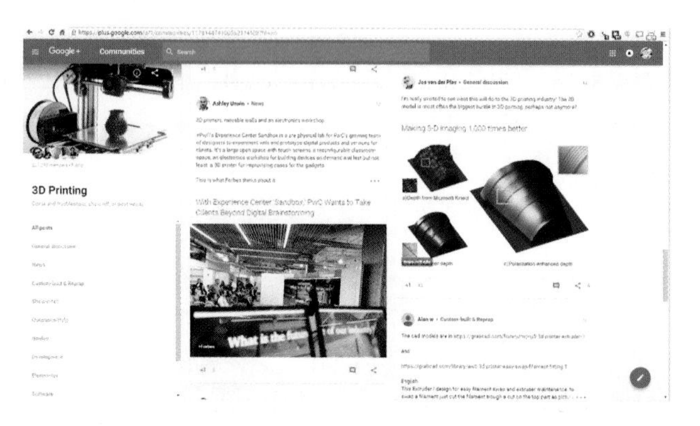

Figure 13-1. *Main landing page of the Google+ 3D printing community*

We Are All Makers, and Companies Will Foster That Even More

3D printing will eventually touch everyone's lives, and companies are starting to notice the potential. Consumers will all be makers and companies are developing products, services, and tools to help them participate even more in this technology.

Hardware stores in the future may not stock all items but rather maintain CAD files of items (even ones that are no longer in production) so you can print them at home or pick up 3D printed copies at the store. You will even be able to modify it if you wish, like the specialized wrench in Figure 13-2.

If your dishwasher breaks, you may be able to download the CAD file for a replacement part via the manufacturers website and 3D print it at home...even at 3 am when stores and repair services are usually closed! This will save you the lengthy wait to get the part delivered in the mail or you having to pick it up at the store.

Companies know you want to save time and money. Companies that offer downloadable CAD files as an option might also provide consumer installation instructions to limit unnecessary service calls. If it's an easy install, consumers will be able to complete the repair themselves.

With 3D printing, you will no longer be bound to the set list of possible choices offered by a manufacturer. You may become the manufacturer yourself!

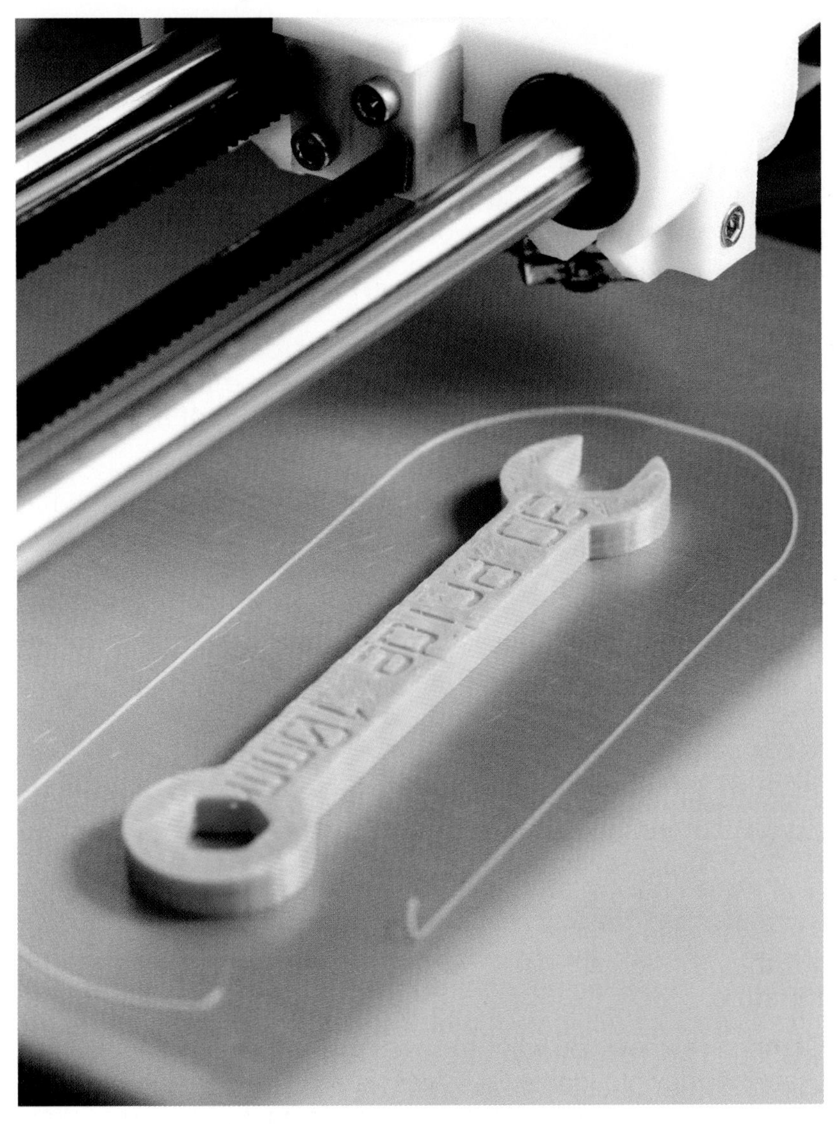

Figure 13-2. *In the future, a tool manufacturer will let you download the CAD file and 3D print it at home or at your local hardware store*

Companies Will Use 3D Printing to Strengthen Their Connection with You

3D printing offers organizations a unique branding opportunity. Mass customization through 3D printing allows for the widespread and relatively inexpensive manufacturing of one-off objects. Brand managers know that most customers want products tailored to them. The ability to create something personal, in partnership with their favorite brand, will strengthen brand loyalty and make customers more involved than just passive buyers.

In the future you will see more examples of branded "one-off" gloves, hearing aids, hats, and more. It will be the new way to monogram your purchases, with just the touch of a button, as shown in Figure 13-3!

Figure 13-3. *User-friendly interfaces will allow for mass customization of 3D printed goods over the Internet*

Companies are motivated to acknowledge your unique needs, interests, and tastes because doing so strengthens their connection with you and increases sales. They will take your design aesthetic and/or actual physical measurements into consideration when they make your (literally *your*) item. Therefore, competitors that don't offer you a 3D printed customized option might come across as uncaring or unwilling to acknowledge your preferences...and that never goes well for sales.

Companies will need to provide a very easy-to-understand user interface when presenting their customers with so much choice to minimize confusion. This combination of challenge and opportunity will certainly change the way companies offer choices and promotional items to their customers in the near future.

Textbook companies will undoubtedly adapt to this exciting new technology, too. Imagine students being able to not only view 3D models online, but also to make these models take physical form, as shown in Figure 13-4.

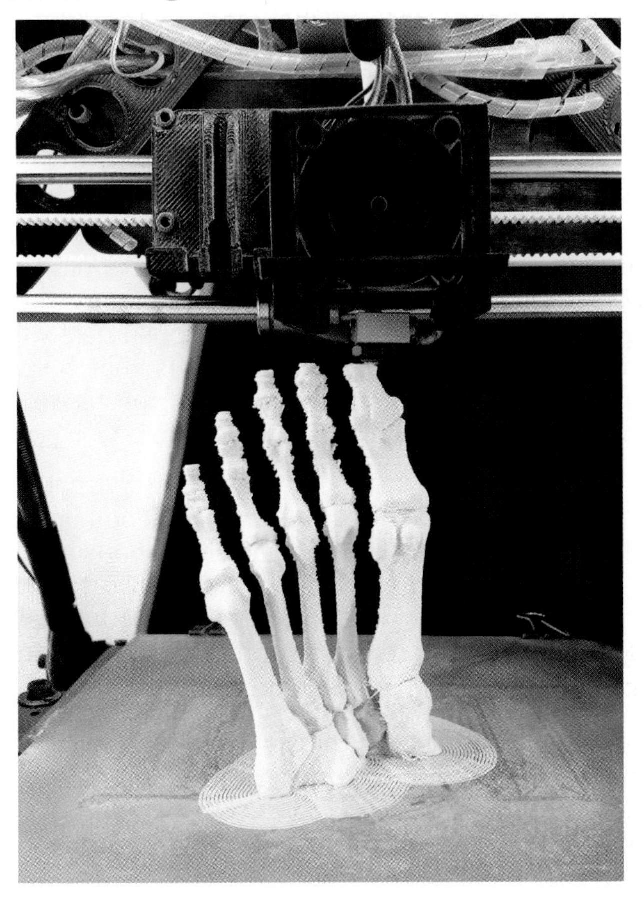

Figure 13-4. *School textbooks will include 3D printable files as part of the curriculum. Shown here is a 3D printed foot chosen by the instructor for use in an anatomy class.*

Organizations Will Increase Profitability and Be More Eco-Friendly

3D printing will offer all organizations (both profit and nonprofit) other benefits than just generating brand loyalty and increased sales. 3D printing will allow organizations to increase their profitability and decrease their environmental footprint. Organizations will be able to make just what they need, as they need it, without the waste normally associated with mass production. Additionally, if a design flaw is found in a product, instant changes can be made at much lower costs than would be the case for retooling the entire production line.

3D printing drastically changes the supply chain by bringing manufacturing closer to the end user and initiating the manufacturing of a product at the moment a customer wants to buy it, not before. Environmentally conscious consumers will rejoice in 3D printing's ability to produce one-offs that help control inventory waste, reduce shipping distances, shorten lead times, and minimize storage needs, as shown in Figure 13-5.

Organizations that stock more CAD files than physical items can reduce the amount of wasted inventory that ends up as landfill and, therefore, become more ecologically responsible. The distinction between shopping online and shopping at a brick-and-mortar location may become less clear as your local garden supply store, clothing store, or even pharmacy begins to include CAD files in its inventory alongside the items on the shelves.

Figure 13-5. *Excess inventory that may be defective, unnecessary or waiting to be sold*

Local Economies Will Benefit from the Commercial Use of 3D Printing

As we mentioned before, 3D printing will not replace mass production but rather will integrate into current manufacturing models. Specialized makers, potentially including you, will have the ability to economically manufacture "low volume" production runs ranging in numbers from 1 to 1,000 units.

With 3D printing, you will see products being made for hyperlocal markets like cities, towns, or even neighborhoods. This hybrid model will help keep more revenue for local manufacturing areas. In addition, local economies will also get the tax revenues associated with the sales transactions.

There are organizations around the world that want to encourage an environment of local manufacturing. One such organization is America Makes (*http://www.americamakes.us*), a premier national accelerator for the 3D printing industry that encourages the growth of US-based manufacturing. A private/public initiative formed by the Obama Administration in 2012, America Makes aims to bring back local jobs and manufacturing to the USA through the application of advanced additive manufacturing (another name for 3D printing).

But job candidates will have to adapt. Local job seekers, wanting local manufacturing jobs, will have to acquire new skills that prepare them to work in this new manufacturing environment. Candidates entering this workforce will have more opportunities if they are proficient in CAD modeling, 3D printing management, and the related technologies.

The Best Is Yet to Come

3D printing is already changing our lives financially, psychologically, socially, and creatively. As the industry and technologies advance, 3D printing will continue to change our lives in ways *we* can't even imagine.

Remember, this technology is still young and is expanding every few months. 3D printing will continue to evolve and strengthen as its ecosystem is deepened by new companies and new applications that emerge in this industry.

As you continue your journey into the world of 3D printing, keep in mind that the most important application may have yet to be discovered. Maybe it will be created by you!

Index

G

H

I

K

L

M

N

O

P

R

S

T

U

V

W

X

About the Authors

Nick Kloski

Nick has earned the respect of his colleagues through hard work, dedication, and vision throughout his more than 16 years in the high-tech industry. Graduating from UC Santa Barbara with an English major, he was hired by Sun Microsystems during the dot-com boom and has since held a number of technical roles at Sun Microsystems and Oracle translating complex technical architectures into understandable ideas.

Nick has a deep understanding of the technical and mechanical side of the 3D printing industry, as well as a strong vision of how the industry has the capability to inspire the world for the better.

Nick cofounded HoneyPoint3D™ (*http://www.honeypoint3d.com*) in 2013 with his wife Liza and now leads HoneyPoint3D's education and events division. HoneyPoint3D™ has won numerous industry awards and is a leader in 3D printing online classes, rapid prototyping, 3D modeling services, 3D scanning services, and industry consultations.

Liza Wallach Kloski

Liza Wallach Kloski is originally from Guadalajara, Mexico. She founded LizaSonia Designs in 2003, a unique upscale jewelry brand and retail store in the Montclair District of Oakland, which wholesaled designs in more than 17 Nordstrom stores, as well as 80 other retail stores.

A graduate of UC Berkeley, Liza has won numerous design and business awards and was the main retail expert in *Entrepreneur* magazine's paperback book *Start Your Own Fashion Accessories Business* (StartUp Series). LizaSonia Designs was a successful jewelry company for over 10 years before Liza's passion turned to the 3D printing industry.

A cofounder of HoneyPoint3D™ (*http://www.honeypoint3d.com*), Liza now leads HoneyPoint3D's rapid prototyping division and manages one of the Bay Area's largest meetup groups on 3D printing, "Bay Area's 3D Printing, 3D Modeling and 3D Scanning Mixers."

As recognized 3D printing experts, Nick and Liza have been interviewed by CNN, RT Television, San Francisco Business Times, KGO Radio, most Bay Area newspapers, and many other international media venues.

Colophon

The image on the cover of *Getting Started with 3D Printing* is a rendered apple that is half wire frame and half 3D printed, © HoneyPoint3D™. The cover fonts are Benton Sans and Soho. The text font is Benton Sans; the heading font is Serifa; and the code font is The Sans Mono.